The Royal Geographical Society

POLAR EXPLORATION

BEAU RIFFENBURGH

ANDRE DEUTSCH

Contents

Introduction	5
Prelude to the Heroic Age	6
Death in the Arctic	8
New Lands in the Far North	10
Scurvy Halts the British	12
The Northwest Passage	14
An American Tragedy	16
Dining on the Dead	18
Nansen Crosses Greenland	20
The Voyage of *Fram*	22
Balloon Towards the Pole	24
Belgica's Antarctic Winter	26
Wintering on the Continent	28
Second *Fram* Expedition	30
Scott of the Antarctic	32
Nordenskjöld's Adversity	34
The Scientists Head South	36
Charcot's Two Voyages	38
The Northwest Passage	40
Shackleton's Farthest South	42
The Great Controversy	44
Scott's Last Expedition	46
Amundsen Attains the Pole	48
Filchner Foiled by the Ice	50
Mawson's Race with Death	52
Shackleton's Endurance	54
The Age of Flight	56
After the Heroic Age	58
Translation	60
Further Reading	61
Index/Credits	62

INTRODUCTION

FRIDTJOF NANSEN, ROBERT FALCON SCOTT, ERNEST SHACKLETON, ROALD AMUNDSEN, ROBERT E PEARY – THESE WERE AMONGST THE GREATEST EXPLORERS OF ANY REGION OF THE WORLD. WHAT IS MOST REMARKABLE ABOUT THESE PARTICULAR MEN, HOWEVER, IS NOT SIMPLY THEIR AMAZING INDIVIDUAL EXPLOITS AND ACCOMPLISHMENTS. RATHER, THAT THEY ARE ONLY A HANDFUL OF A MUCH LARGER NUMBER OF EXTRAORDINARY MEN WHO IN ONE BRIEF, INTENSE PERIOD COMBINED TO CONQUER THE MYSTERIOUS, ICY WASTES OF BOTH THE ARCTIC AND THE ANTARCTIC SO THAT THEY WOULD NO LONGER BE BLANK SPACES ON THE MAP. IN DOING SO, THEY PROVED BEYOND DOUBT THAT THERE WAS NOWHERE ON EARTH MAN COULD NOT TREAD AND NO GEOGRAPHICAL FEAT HE COULD NOT ACCOMPLISH.

For centuries the polar regions had represented some of the most terrible and dangerous aspects of nature. They had fiercely guarded their secrets, and countless men had been doomed to watery or icy graves in these forlorn regions. But then, in less than 50 years – the last quarter of the nineteenth century and the first two decades of the twentieth – explorers from Britain, the United States, Norway, Sweden, Germany, France, Russia, Italy, Austria, Australia and Japan conquered nature's last, closely guarded bastions and comprehensively opened the ends of the Earth to mankind's knowledge.

Polar Exploration is the story of those men, and of the time in which they lived and – in some cases – died: the "Heroic Age" of polar exploration. Such a period, with so many great explorers launching so many expeditions in such geographical and temporal proximity, was unprecedented, as were their countless achievements. In a seeming flurry, they discovered many of the world's last known islands, first navigated the Northwest Passage and the Northeast Passage, first attained the North Pole and the South Pole, first crossed the Arctic Basin and first flew airships and aeroplanes in the Arctic and Antarctic. But they achieved much more than mere geographical discoveries, and the tales of their celebrated successes and glorious failures still quicken pulses and stir hearts.

That is just how many of these explorers would have wanted it, because, although they were driven in the Heroic Age, as throughout history, by a variety of goals, fame was certainly one of the strongest. As Peary wrote uncompromisingly to his mother: "I *must* have fame, and I cannot reconcile myself to years of commonplace drudgery…I want my fame *now*."

Peary's arch-rival Frederick Cook was also looking for fame at any cost, as were others such as Carsten Borchgrevink and Richard E Byrd. The fortune that can accompany fame was another hoped-for reward of many explorers, and no one seemed more intent on earning such wealth than Shackleton, who had more than a little of both the gambler and the buccaneer in his personality.

Exploration in the polar regions was also driven by numerous other factors. The desire for the "conquest of the world" – that is, to prove man could overcome any obstacles set by the natural world – was a crucial underlying component in the nineteenth century. National pride was key to some explorers; Nansen, for example, believed that his triumphs would gain credibility for Norway in her struggles for independence, while Scott hoped to show that Britons had lost none of the vigour and courage that had allowed them to establish the greatest empire on Earth. Commerce also played a major role: the voyages undertaken by Joseph Wiggins into the Kara Sea hoped to prove that commercial shipping from Europe to Siberia was practical, a concept also of significance in Adolf Erik Nordenskiöld's navigation of the Northeast Passage.

Scientific research became increasingly important. From a time when science on Arctic expeditions was carried out as virtually an aside by Royal Navy officers, it grew to where it was the *raison d'être* for Nansen's Arctic expeditions and for the Antarctic efforts led by Erich von Drygalski, W S Bruce and Douglas Mawson.

As such disparate goals indicate, the Heroic Age featured incredible diversity, including men of many nationalities, intellectual persuasions and physical abilities. These differences meant that the strengths and successes of explorers could be demonstrated, but so could their weaknesses and failures. Thus, this book contains tales of Shackleton's open-boat journey, Mawson's solo trek across the vast reaches of the Polar Plateau, the perfection of Amundsen's efforts both north and south and the noble actions of Captain Oates. But it also narrates stories less unquestioningly positive: a possible poisoning in the High Arctic; the tragic disappearance of three men in a balloon; the madness brought about by the first Antarctic wintering; the horrors of cannibalism and false claims, near misses, great controversies and that most-feared killer: scurvy. As the following pages show, these were all an integral part of the Heroic Age of polar exploration.

Beau Riffenburgh

POLAR 1818–1859
PRELUDE TO THE HEROIC AGE

AFTER THE CONCLUSION OF THE NAPOLEONIC WARS, SIR JOHN BARROW, THE SECOND SECRETARY OF THE ADMIRALTY, PUT FORWARD A PLAN TO KEEP SOME OF THE OFFICERS AND MEN OF THE ROYAL NAVY ACTIVE, RATHER THAN ALLOWING THEM TO GO INTO SEMI-RETIREMENT ON HALF-PAY. HE PROPOSED SENDING OUT EXPLORING EXPEDITIONS AIMED AT THE CENTURIES-OLD GOAL OF FINDING THE NORTHWEST PASSAGE, A SEA ROUTE TO ASIA THROUGH THE MYSTERIOUS WATERWAYS OF THE ARCTIC REACHES OF BRITISH NORTH AMERICA.

ABOVE The expedition under John Ross in 1818 marked the beginning of new efforts by the Admiralty to navigate the Northwest Passage via the three sounds (Smith, Jones and Lancaster), exiting Baffin Bay.

In the following decades, a series of expeditions led by John Ross, John Franklin, William Edward Parry and George Back, amongst others, followed up on earlier discoveries by Samuel Hearne and Alexander Mackenzie and explored little-known regions of the Canadian archipelago and mainland, although none was able to complete the Passage. On one of these expeditions, Ross' nephew, James Clark Ross, became the first man to reach the North Magnetic Pole on 1 June 1831. This increase in geographical discovery was accompanied by a growth of interest in science. Frederick Beechey, Edward Sabine and the famed whaler William Scoresby Jr were just a handful of the men advancing scientific knowledge on Arctic expeditions.

Antarctic ventures likewise flourished in the early decades of the nineteenth century, with both geographical exploration and scientific investigation playing major roles. In 1819–1821, Fabian von Bellingshausen of the Russian Navy became the second commander (after Captain James Cook) to circumnavigate the Earth in the far south, helping further delineate the placement of the Antarctic continent, of which he almost certainly made the first sighting. In the late 1830s, three major national exploring expeditions set out for the Antarctic: the French under Jules-Sébastien-César Dumont d'Urville, the United States Exploring Expedition under Charles Wilkes and a Royal Navy expedition commanded by James Clark Ross. Although each charted areas of coastline, and Ross discovered the Ross Sea, Ross Island, and the Great Ice Barrier – now known as the Ross Ice Shelf – the icy conditions which endangered the three expeditions did not encourage further costly exploration in that region. By the middle of the century, the focus of exploration had shifted away from the Antarctic, where it would not return until the 1890s.

ABOVE A snuff box made from the wood of HMS *Hecla*, once commanded by W E Parry. The cover includes the arms of the city of Winchester, of which Parry was made a free man.

Meanwhile, in 1845, Sir John Franklin left with two ships on the most lavishly equipped Northwest Passage effort ever launched. But within a few years, it was realized that the expedition had simply disappeared into the uncharted wastes of the high Canadian Arctic. Beginning in 1847, a series of expeditions left to look for it. Some of these were sent by the Royal Navy or the Hudson's Bay Company, others sponsored by Lady Franklin and aided by public subscription, while several American expeditions received financing from the wealthy merchant Henry Grinnell.

In 1850, on one of these expeditions, Robert McClure, in *Investigator*, approached the search area from Bering Strait. However, the ship became beset in the ice, and it was not until 1853 that she was finally abandoned and the crew trekked west over the ice to join the *Resolute*, which had entered the heart of the archipelago from the east as part of another expedition searching for Franklin. When McClure's party sailed for home, it became the first to complete the Northwest Passage, although the feat would not be accomplished

LEFT Fabian von Bellingshausen made a pioneering voyage to the Antarctic, but little was known outside of Russia about his discoveries until the 1940s, when Frank Debenham edited Bellingshausen's translated account.

BELOW John Rae of the Hudson's Bay Company had already led two Franklin search expeditions when, in the spring of 1854, he heard tales about the fate of Franklin's party, which would cause an uproar in England.

ENCLOSURES

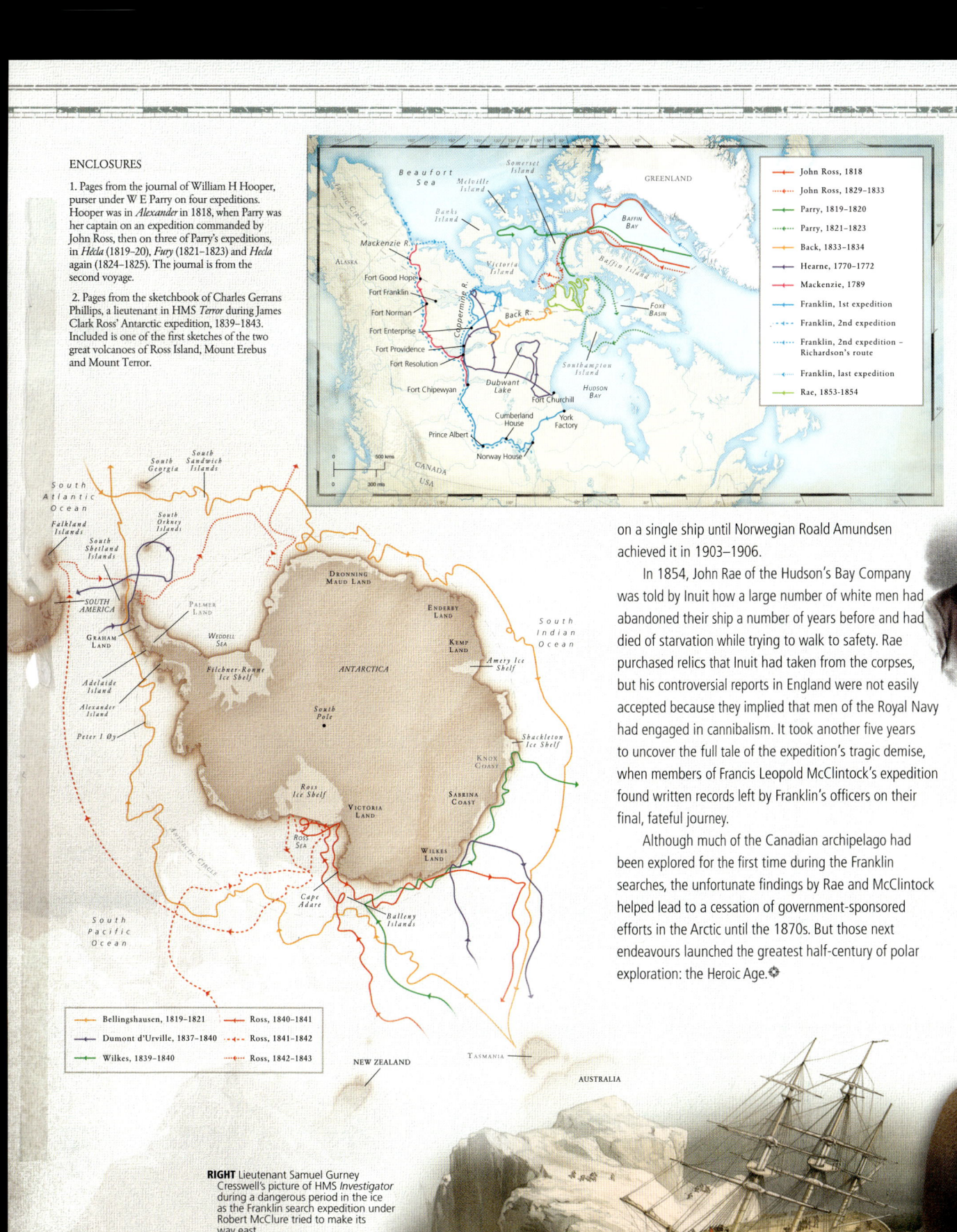

1. Pages from the journal of William H Hooper, purser under W E Parry on four expeditions. Hooper was in *Alexander* in 1818, when Parry was her captain on an expedition commanded by John Ross, then on three of Parry's expeditions, in *Hecla* (1819–20), *Fury* (1821–1823) and *Hecla* again (1824–1825). The journal is from the second voyage.

2. Pages from the sketchbook of Charles Gerrans Phillips, a lieutenant in HMS *Terror* during James Clark Ross' Antarctic expedition, 1839–1843. Included is one of the first sketches of the two great volcanoes of Ross Island, Mount Erebus and Mount Terror.

- John Ross, 1818
- John Ross, 1829–1833
- Parry, 1819–1820
- Parry, 1821–1823
- Back, 1833–1834
- Hearne, 1770–1772
- Mackenzie, 1789
- Franklin, 1st expedition
- Franklin, 2nd expedition
- Franklin, 2nd expedition – Richardson's route
- Franklin, last expedition
- Rae, 1853–1854

- Bellingshausen, 1819–1821
- Dumont d'Urville, 1837–1840
- Wilkes, 1839–1840
- Ross, 1840–1841
- Ross, 1841–1842
- Ross, 1842–1843

on a single ship until Norwegian Roald Amundsen achieved it in 1903–1906.

In 1854, John Rae of the Hudson's Bay Company was told by Inuit how a large number of white men had abandoned their ship a number of years before and had died of starvation while trying to walk to safety. Rae purchased relics that Inuit had taken from the corpses, but his controversial reports in England were not easily accepted because they implied that men of the Royal Navy had engaged in cannibalism. It took another five years to uncover the full tale of the expedition's tragic demise, when members of Francis Leopold McClintock's expedition found written records left by Franklin's officers on their final, fateful journey.

Although much of the Canadian archipelago had been explored for the first time during the Franklin searches, the unfortunate findings by Rae and McClintock helped lead to a cessation of government-sponsored efforts in the Arctic until the 1870s. But those next endeavours launched the greatest half-century of polar exploration: the Heroic Age.

RIGHT Lieutenant Samuel Gurney Cresswell's picture of HMS *Investigator* during a dangerous period in the ice as the Franklin search expedition under Robert McClure tried to make its way east.

THE FIRST TOURIST IN THE FAR NORTH

Tourism in the polar regions is big business today. Its beginnings can be traced back to 1856 when Frederick Hamilton-Temple-Blackwood, the Marquis of Dufferin and Ava, organized a cruise of northern seas in his yacht, *Foam*. Dufferin first sailed to Iceland and Jan Mayen, making the first recorded visit to the latter island in almost 40 years. He continued to Norway and then to the west coast of Spitsbergen. This was the first known voyage to the Svalbard archipelago undertaken solely for pleasure, and it inspired numerous other wealthy adventurers to sail north.

LEFT After his Arctic voyage, Lord Dufferin became a highly successful diplomat. Among other positions, he served as Governor General of Canada and then Viceroy of India.

HUNTING FOR SPORT IN THE ARCTIC

One of the people inspired by Dufferin was the hunter James Lamont, who in 1858 and again in 1859 sailed into Svalbard waters to hunt for walruses, seals, polar bears, reindeer and any other animals he could find. His subsequent book about the expeditions made Lamont a well-known figure, as in Victorian times big-game hunting was considered to be a manly sport. After a decade turning his hand to hunting in other areas of the world, Lamont returned to the north, leading three private hunting ventures to the Barents Sea and Novaya Zemlya.

LEFT Although hunting for big game was already extremely fashionable with Europeans, James Lamont helped popularise the high Arctic as a destination for such "sport".

ARCTIC 1871–1873

DEATH IN THE ARCTIC

In 1869, shortly after returning from a small, private expedition to the Canadian Arctic, Charles Francis Hall went to the United States Congress to appeal for funding for an attempt on the North Pole via Smith Sound, between Greenland and Ellesmere Island. Remarkably, Congress approved his request, and in the summer of 1871 the expedition ship *Polaris* headed north, with Hall as commander, Sidney Buddington as master and German physician Emil Bessels as chief scientist. From the start, however, the party split into factions, as Hall, with no experience of command, fell out first with Bessels and the sizeable German contingent, and then with Buddington, who evinced a lack of willingness to carry out the required exploring programme.

Polaris made rapid progress north, going through Smith Sound, Kennedy Channel and Hall Basin before being brought up by heavy ice at 82°11' N, the farthest north that had then been reached by a ship. With Buddington refusing to proceed, Hall retreated to find winter quarters. These were established at 81°38' N on the Greenland shore, at what Hall named "Thank God Harbour", where the men built an observatory and began a series of local surveys.

In September, Hall started to send out exploratory sledge parties, and he sledged north himself for two weeks in October, surveying the inlets of Hall Basin. But upon his return, he almost immediately began suffering from an intense fever. Soon, he became delirious and complained of being poisoned. He died on 8 November and was buried ashore. With Hall's death, Buddington assumed command, but the expedition's discipline disappeared and most of the crew lost any enthusiasm for making an attempt on the North Pole.

In the spring of 1872, assistant navigator George Tyson and mate H C Chester led boat parties into the surrounding areas but did not travel far, as Buddington had stated categorically that he would depart as soon as the ship was free, regardless of whether any parties remained in the field. *Polaris* escaped from the ice and turned towards home in August, but was quickly again beset and drifted aimlessly south. In mid-October, with the ship seemingly in immediate danger of being crushed by the ice, preparations were made to abandon her, and stores, equipment and three boats were put on to a large ice floe. With 19 men on the ice, and only 14 still aboard, the floe suddenly broke away and drifted out of the reach of the ship, which continued helplessly south.

For the next six months the men on the ice floe, under the command of Tyson, drifted slowly south. They survived

ABOVE Inuit fur boots obtained by W E Parry on his second expedition (1821–1823). These were similar to those that would have been worn by the Inuit with whom Hall lived on his first two expeditions.

TOP The only known photograph of Charles Francis Hall, taken in Washington in 1870 with three staunch supporters. Pictured from left: Colonel James Lupton, Major T H Stanton, Hall and Penn Clarke.

LEFT Emil Bessels, the surgeon and chief of scientific staff aboard *Polaris*. Hall was worried that Bessels was poisoning him, and more than a century later some experts still consider this likely.

ENCLOSURE
Part of a book of songs originally belonging to Emil Bessels. After being left behind in the Arctic, it was discovered five years later by H C Hart, naturalist from *Discovery* on the British Arctic Expedition.

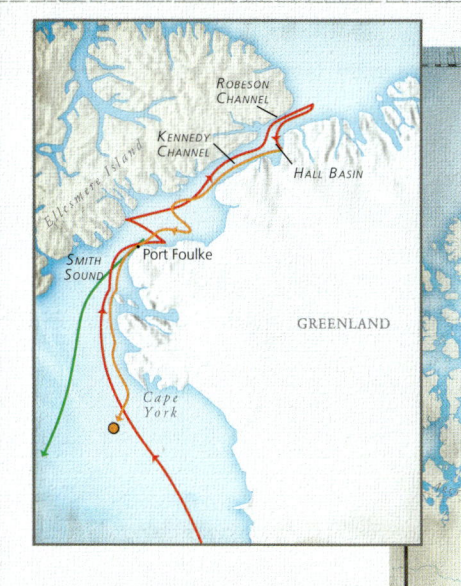

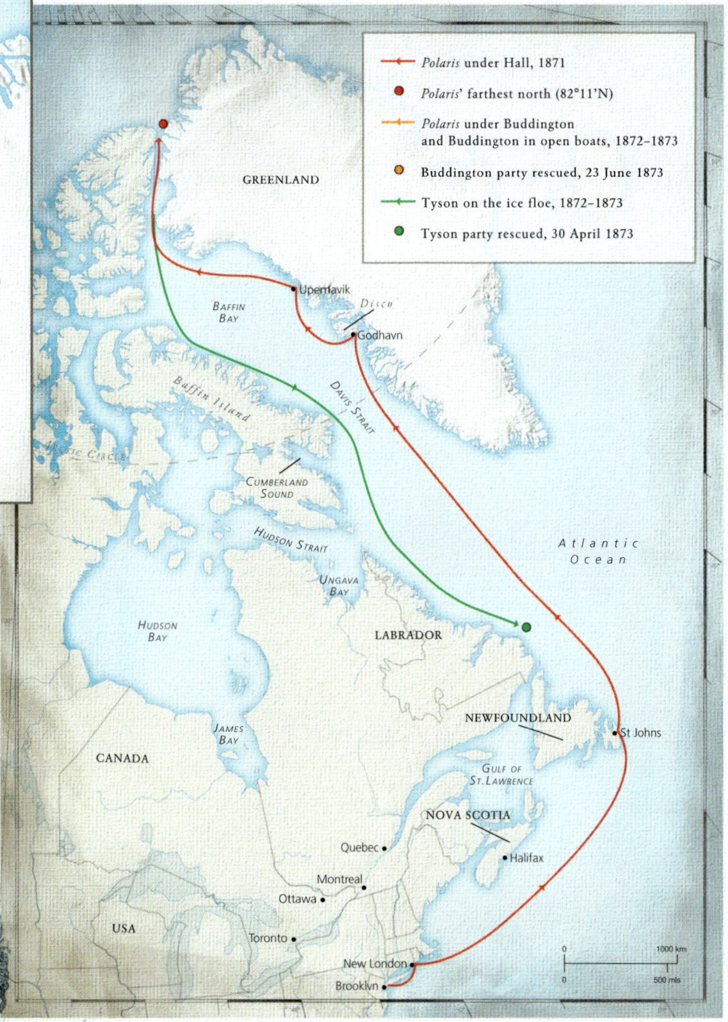

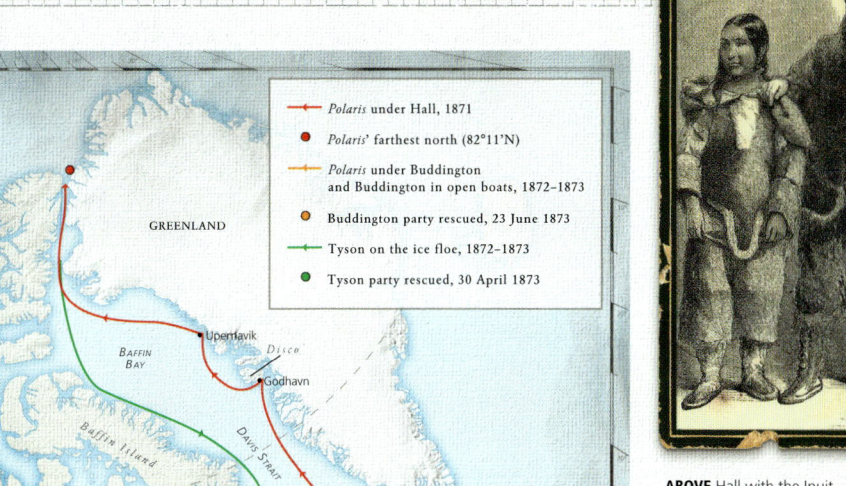

BELOW George Tyson, the senior officer during the remarkable drift on the ice floe. The next year, Tyson served as ice master on USS *Tigress* in her unsuccessful attempt to relieve Adolphus Greely's party.

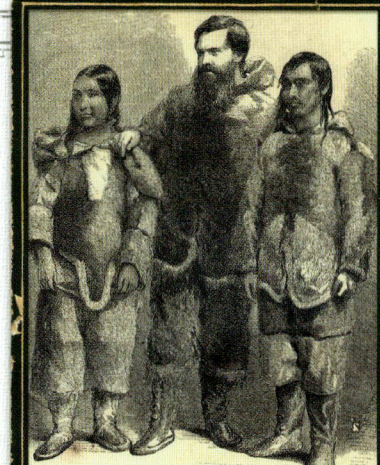

ABOVE Hall with the Inuit couple Tookoolito and Ebierbing, his companions on his first two expeditions

HALL'S EARLIER EXPEDITIONS

Hall was a newspaperman in Cincinnati, Ohio, but in 1860 his obsession with the Franklin tragedy led him to the Arctic to search for survivors of that expedition. For two years, he lived with the Inuit of Frobisher Bay on Baffin Island, adopting many aspects of their way of life. In 1864, he went back to the Canadian Arctic, where in a five-year period in the company of two Inuit he sledged some 6,500 kilometres (4,000 miles), exploring, mapping and finding numerous Franklin relics, including the skeleton of Lieutenant Le Vesconte of Franklin's ship *Erebus*.

THE EXHUMATION OF HALL

After the naval board of inquiry ruled on Hall's death, the subject did not attract further interest for almost a century. But in 1968, Chauncey Loomis of Dartmouth College, who was writing a biography of Hall, exhumed the explorer's body to take hair and fingernail samples. Analysis of these showed that Hall had most likely died from arsenic poisoning. Although in recent years both Buddington and Bessels have been accused of poisoning their leader, it is impossible to tell if he actually was murdered, died from self-administering the poison, or perished from natural causes, such as a stroke.

the winter with its −40°C (−40°F) temperatures only through the knowledge and hunting skills of the Inuit members of the party, who built igloos and killed polar bears and seals. They were finally rescued on 30 April 1873 off the coast of Labrador, having drifted some 3,000 kilometres (1,850 miles) in 197 days and seen the once large floe slowly reduce to only about 68 by 92 metres (75 by 100 yards).

Meanwhile, *Polaris* finally broke free of the ice, but Buddington was forced to take her to nearby Foulke Fjord in Smith Sound, owing to severe leaks. The party spent the winter there in a house built from the timbers of the ship, and in June 1873 set off south in boats. Three weeks later, they were picked up by a whaler and taken to Dundee, from where they all eventually returned to New York.

An official naval board of inquiry subsequently investigated the death of Hall and the major loss of discipline that followed it. With details that shocked the public, the conflicts among the officers before and after their commander's death, the rampant drunkenness during the winter and Buddington's failure to go back to the aid of the men left on the ice floe were brought to light. However, the board caused less sensation with its ruling that Hall had simply died of apoplexy.

ABOVE The procession during Hall's funeral in northern Greenland on 10 November 1872. Owing to the darkness throughout much of the day, the men had to find their way to the gravesite with lanterns.

ARCTIC 1872–1874
NEW LANDS IN THE FAR NORTH

IN 1872, ENCOURAGED BY FAMED GEOGRAPHER AUGUST PETERMANN'S THEORIES OF AN OPEN POLAR SEA — A WARM, ICE-FREE REGION SURROUNDING THE NORTH POLE INSIDE A THICK RING OF ICE — THE AUSTRIAN GOVERNMENT SENT OUT A MAJOR EXPEDITION: THE AUSTRO-HUNGARIAN EXPLORING EXPEDITION. ITS GENERAL GOAL WAS TO EXPLORE NORTHWARDS INTO THE ARCTIC OCEAN FROM NEAR NOVAYA ZEMLYA, WHERE PETERMANN WAS THEN PREDICTING THE OPEN POLAR SEA WOULD BE FOUND. HOWEVER, ITS SPECIFIC PLANS WERE VAGUE: NO WINTERING SITE WAS DECIDED UPON, AND BOTH THE NAVIGATION OF THE NORTHEAST PASSAGE AND AN ATTEMPT ON THE POLE WERE CONSIDERED POSSIBLE.

Sailing under joint leaders, Karl Weyprecht (for sea operations) and Julius Payer (for land operations), the expedition's ship *Tegetthoff* crossed the Barents Sea and proceeded up the west coast of Novaya Zemlya. But within a day of leaving the supply ship, the *Tegetthoff* became beset in the ice and began to drift erratically back and forth across stretches of the Arctic Ocean. All efforts to force open a channel with ice saws and picks were in vain, and the crew nervously settled in for the long polar winter, in which the continual darkness, the crashing of the ice, the creaking of the ship and the extreme cold were almost more than they could stand.

By late summer 1873, still held in the ice, the ship had reached 79°43′ N, 59°33′ E. Suddenly, on 30 August, the fog lifted, and before their eyes appeared the outline of a mountainous region never before seen. The discovery of new land revived the men's spirits, as did naming it "Kaiser Franz Josefs Land" (later known as Franz Josef Land). However, at the time they were still some 46 kilometres (28 miles) away from it, and the ice and weather conditions prevented any long exploratory trips from the ship. It was not until November, after the sun had set for a second polar winter, that Payer was able to lead a sledge party to what they named Wilczek Island, after the expedition's most generous sponsor. No more sledge trips were possible in the darkness, and the ship remained locked in the ice next to Wilczek Island. Despite being able to kill polar bears for fresh meat, most of the crew suffered from a variety of diseases, and the engineer, Otto Krisch, died of scurvy and was buried ashore.

In March 1874, Payer led the first sledge journey to explore the unknown lands. Heading west for several days, the party discovered Hall Island. A much longer effort began later that month, when Payer, six men and three dogs headed north along Austria Sound. Three of the men were soon sent to survey Hohenlohe Island while the remainder pushed north. Seventeen days after leaving the ship, Payer's party reached its northernmost point — 82°5′N — at Cape Fligely on Rudolf Island (named for the Crown Prince of Austria-Hungary). To the north, he thought he could see another island, which he called Petermann Land. However, this later proved to be non-existent; the party had, in fact, reached the northern extreme of the archipelago. The return was fraught with concern that the ship might have drifted away, but in late April they found it still held by the ice off Wilczek Island.

LEFT Julius von Payer, co-leader of the Austro-Hungarian Exploring Expedition. Payer had previously served as cartographer on Karl Koldewey's *Germania* (1869–1870), and had then been co-leader (with Weyprecht) on a reconnaissance expedition in 1871.

ABOVE Travelling through the unknown archipelago required that the boats be used both on waterways and as sledges over the ice. This painting is by Adolf Obermüllner, based on a sketch by Payer.

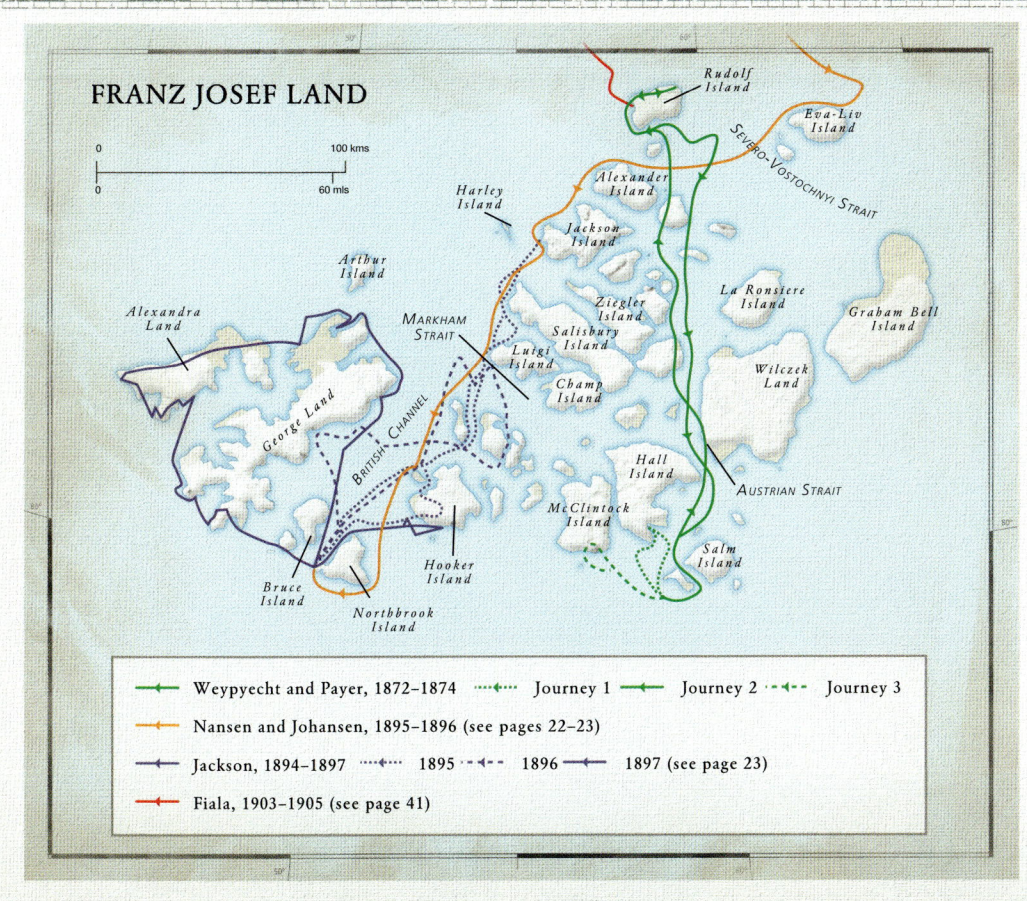

FRANZ JOSEF LAND

Map legend:
- Weypyecht and Payer, 1872–1874 — Journey 1 — Journey 2 — Journey 3
- Nansen and Johansen, 1895–1896 (see pages 22–23)
- Jackson, 1894–1897 — 1895 — 1896 — 1897 (see page 23)
- Fiala, 1903–1905 (see page 41)

RIGHT Despite his controversial ideas, Petermann was respected throughout the world, and numerous expeditions were launched to test his theories.

BELOW The members of the expedition in the process of abandoning *Tegetthoff*. They left hauling four boats, and it was just short of three months before they could launch them into open sea.

In mid-May, after a final exploratory sledge trip discovered McClintock Island, *Tegetthoff* was abandoned. Taking the scientific results, but leaving behind much of the natural-history collection, the 23 men retreated south, hauling open boats over the ice. But after two months of gruelling work, they suddenly spied *Tegetthoff* in the distance. As it turned out, the head-winds into which they had been struggling had blown the ice back north, and they had made almost no true progress south. Most of the men wanted to return to the ship, but Weyprecht persuaded them that that they must continue on their journey south. Shortly thereafter, the wind direction changed, and in mid-August they launched their boats into the open sea.

In the following days, the party rowed and sailed their boats to the west coast of Novaya Zemlya, where, after proceeding about half-way down the southern island, they were rescued by two Russian schooners.

AUGUST PETERMANN

A remarkably talented cartographer, Petermann was so highly regarded during a decade living in Britain that he was named official physical geographer to Queen Victoria. He gained even greater international stature when he moved to Gotha in Germany and founded the influential journal *Petermann's Geographische Mitteilungen*, which reflected his interests in scientific geography and exploration. Petermann was consulted regularly about exploration of central Africa and the Arctic. A strong proponent of the notion of an open polar sea, his theories about ocean currents and their relation to such a body of water influenced numerous expeditions to the Arctic.

KOLDEWEY'S SEARCH FOR THE OPEN POLAR SEA

The Austro-Hungarian Exploring Expedition was not the first Arctic effort to be based on Petermann's theories. In 1869, a German expedition under Karl Koldewey headed to northeast Greenland, where Petermann predicted the open polar sea would be found. The two ships became separated and one, *Hansa*, was crushed in the ice; her crew wintered there before proceeding to a Danish settlement. Koldewey took the other ship, *Germania*, up the east coast until halted by heavy ice. Further efforts the next spring proved equally unsuccessful, leading Petermann to decide that the open polar sea might instead best be reached between Svalbard and Novaya Zemlya.

RIGHT Karl Koldewey, who was selected by Petermann to lead the expedition, which was funded by the geographer.

ARCTIC 1875–1876: SCURVY HALTS THE BRITISH

In 1864, Sherard Osborn, who had twice commanded a ship during the Franklin searches, read a paper to the Royal Geographical Society pressing for a renewal of British polar exploration and research. For another decade, he continued his outspoken campaign, and in 1874 his efforts paid dividends, as Benjamin Disraeli's British government agreed on a national effort to reach the North Pole by way of Smith Sound, between Ellesmere Island and Greenland. Among the first steps were to recall George Strong Nares, who was captain of the research ship *Challenger* in the far south, and to place him in command of what was officially named the British Arctic Expedition.

BELOW George Strong Nares first visited the Arctic as second officer aboard HMS *Resolute* during a Franklin search expedition. He was knighted after the British Arctic Expedition and retired in 1892 as a vice-admiral.

In the summer of 1875, two ships, HMS *Alert* and HMS *Discovery*, having been fitted out for a wide range of scientific studies, sailed north. They made their way through Kane Basin and Kennedy Channel and reached Lady Franklin Bay on Ellesmere Island (on 25 August), where *Discovery* proceeded to winter. Nares, on *Alert*, continued to Floeberg Beach on Ellesmere at 82°28′ N, then a record for the highest latitude reached by a ship. Ostensibly, the men's health remained good through the winter, although in reality their store of vitamin C was becoming depleted.

In April 1876, three major sledge parties set out. One, under Albert Hastings Markham, was assigned to head towards the Pole; the party followed the coast of Ellesmere Island to Cape Joseph Henry and then turned due north. However, Markham and the 16 other men in his party were faced with back-breaking work, man-hauling two extremely heavy sledges over hummocky sea ice in temperatures dropping to -36°C (-33°F). Soon they slowed to a crawl, as virtually all the party began to show symptoms of severe scurvy. By the time they reached 83°20′ 26″ N on 12 May 1876, the farthest north that had ever been attained, only a handful of them were still fit to pull the sledges. Markham wisely turned back, but the return journey was a nightmare owing to the men's weakness, which made hauling the sledges extremely difficult. At the beginning of June, Markham sent Alfred Parr ahead for help, and on 10 June, after one man had died, they were met by a relief party.

Meanwhile, two other field parties had also suffered dreadfully. One, under Pelham Aldrich, had traced the northern coast of Ellesmere Island, discovering Ward Hunt Island and many bays and capes, including Cape Columbia, the most northerly land point in the Canadian Arctic. They also were badly affected with scurvy, but managed to return without loss of life. The third party, led by Lewis Beaumont, explored the north coast of Greenland, but virtually all the

ABOVE HMS *Alert* in the ice. During the winter of 1875–1876, Nares followed the same practices for ensuring his crew's comfort, entertainment and education – such as conducting exercises, running a school and hosting amateur theatricals – as had been established during the Franklin searches.

LEFT An artist's impression of the conditions through which the men of the British Arctic Expedition had to sledge. Man-hauling was once described as the hardest work ever done by free men.

BELOW The chemical laboratory on HMS *Challenger*. It was only one of several labs aboard the ship.

NARES AND THE *CHALLENGER* EXPEDITION

Before taking command of the British Arctic Expedition, Nares had been captain of HMS *Challenger* during the most extensive scientific examination of the world's oceans ever conducted. In 1872, having been converted for oceanographic studies, *Challenger* began her cruise, which would last four years, circumnavigate the Earth and visit virtually every maritime region of the planet. Under the direction of Charles Wyville Thomson, the scientific team charted some 363 million square kilometres (140 million square miles) of ocean floor, discovered a multitude of new species and produced scientific reports that filled 50 volumes.

SCURVY

After the increase in long-distance voyages in the fifteenth century, scurvy became the greatest menace to the health of sailors. Caused by a deficiency of vitamin C, scurvy first appears as a swelling of the gums and loosening of the teeth. Untreated it can lead to death. Through the centuries, various cures were found – such as eating fresh fruit or vegetables – but such discoveries were then lost because of numerous confounding factors. Scurvy was also a problem on expeditions to the polar regions, as the tinned foods regularly used had lost most of their vitamin C.

men became so ill with scurvy that they could not return to the ship on their own and were forced to wait for relief. Two died before they were rescued.

Meanwhile, a number of the men who had remained with the ships had also suffered from scurvy. Nares was perplexed, because he had taken the usual anti-scorbutic precaution of providing lime juice for his men, but, unbeknownst to them at the time, the juice used had lost most of its vitamin C while waiting to be prepared for the expedition and in the subsequent heating and bottling procedures. Trying to avoid further loss of life, Nares terminated the expedition, and in August the two ships sailed for Britain.

The success of the expedition was judged harshly in some quarters, because of the early return to England, the failure of Markham's party to make an approach to the Pole and the problems with scurvy. However, it did make significant geographical discoveries on the other sledging journeys and returned with new scientific findings. A subsequent Parliamentary enquiry into the scurvy, however, did little to explain the reasons why the expedition had been so plagued, but rather concluded in part that lime juice must indeed not be the answer to scurvy. Thus, the expedition had, in a sense, contributed to yet more confusion over the causes of scurvy. ✤

LEFT A total of 35 dogs and two Greenlandic dog handlers were brought on the expedition. However, they never played a major part in the main sledging trips, which were carried out the old-fashioned way: by man-hauling.

ENCLOSURES

1. An autographed menu card from a dinner given to welcome back Captain George Strong Nares and the men of the British Arctic Expedition. The dinner was held in December 1876, a month after the ships reached Portsmouth.

2. Pages from the diary of H W Feilden, naturalist on *Alert* on the British Arctic Expedition. An early page shows the mandatory winter routine that both ships followed.

BELOW A drawing from the expedition showing the crew receiving the regular ration of lime juice to fight scurvy. It was served at five bells of the forenoon.

13

ARCTIC 1878–1880
THE NORTHEAST PASSAGE

LIKE THE MORE FAMOUS NORTHWEST PASSAGE, THE NORTHEAST PASSAGE – THE WATERWAY NORTH OF RUSSIA AND SIBERIA – HAD LONG ATTRACTED EXPLORERS WHO HOPED TO REACH THE RICHES OF THE FAR EAST. BUT ALTHOUGH EXPEDITIONS TO CONQUER IT HAD BEEN SENT OUT AS EARLY AS THE SIXTEENTH CENTURY, NONE HAD BEEN SUCCESSFUL, AND THE ICY CONDITIONS IN THE BARENTS AND KARA SEAS PREVENTED MOST EARLY EXPLORERS EVEN FROM SAILING THROUGH MAJOR PARTS OF IT, MUCH LESS THE ENTIRE PASSAGE.

ABOVE A famed painting of Nordenskiöld with *Vega* in the ice of the Northeast Passage. The explorer's celebrated return convinced many young Swedes to try to follow in his footsteps, including the great Sven Hedin.

RIGHT Aleksandr Sibiryakov, a Russian gold-mining magnate who gave financial backing to both Joseph Wiggins and Nordenskiöld with the hope of opening up trading between Siberia and Western Europe.

In 1874, however, a British merchant named Joseph Wiggins led a venture to demonstrate the feasibility of reaching the Ob' and Yenisey rivers through the Kara Sea, with the goal of developing trade between Western Europe and Siberia. His success in getting to the Ob' not only led him to make an extensive series of expeditions along this route, but encouraged others to try the same.

The most important of those who followed in Wiggins' wake was the Finnish-born Swedish geologist Adolf Erik Nordenskiöld, who had participated in six Swedish expeditions to Svalbard or Greenland before turning his eyes east. With the financial backing of Göteborg businessman Oscar Dickson, Nordenskiöld sailed to the Yenisey in 1875 and again the next year to prove that commercial shipping from Europe to Siberia could be practical.

Navigating the Northeast Passage was Nordenskiöld's logical next step, because it allowed him to combine the commercial interests of three men willing to sponsor him – Dickson, the Russian merchant Aleksandr Sibiryakov and King Oscar II of Sweden – with his own desire to collect and analyze new scientific data. Such information would allow him to understand the Arctic: how it was formed, what lived in the little-known area east of the Taymyr Peninsula and what secrets of science could be explained there. Successfully completing the Passage would also be a significant symbolic act of conquest over nature, which would also reflect positively on his adopted Sweden.

In July 1878, Nordenskiöld set out from Tromsø in the ship *Vega*, accompanied by the merchant vessel *Lena*, which had been fitted out by Sibiryakov. The ships made a surprisingly easy crossing of the Barents Sea and in August sailed past Novaya Zemlya and into the Kara Sea. Despite much fog, the men observed many uncharted islands, and in mid-August the two ships reached Cape Chelyuskin, the northernmost point of Asia. The mass of ice that usually surrounds that cape makes it the most formidable barrier to shipping in the Northeast Passage, but *Vega* and *Lena* succeeded in becoming the first ships on record to round it.

Nordenskiöld had hoped to sail directly across the Laptev Sea, but heavy ice and persistent fog forced him to

ABOVE *Vega* as she supposedly dashed past the spouting whales of the Barents Sea en route to navigating the Northeast Passage. Note the difference in the artists' conceptions between this and the painting on the left.

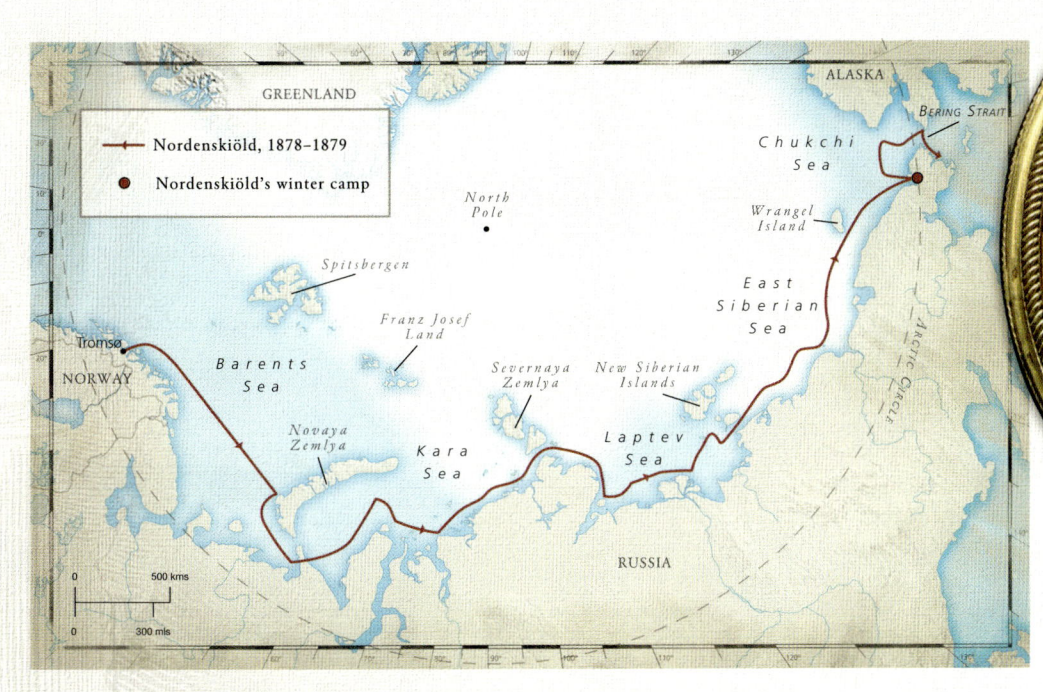

BELOW Christmas 1878 aboard *Vega*, while she was held in the ice in Kolyuchin Bay. Despite being only two days open sailing from Bering Strait, she was trapped for 264 days. Nordenskiöld is sitting at the table in the centre.

ABOVE Otto Torell was an early proponent of scientific study of the Arctic as a specific goal, not just an addendum to geographical investigations.

stay close to the coast. Even this had benefits, as he made the first detailed mapping of the coast of the Taymyr Peninsula. In late August, the ships reached the Lena Delta, and *Lena* headed south to ascend the river. *Vega* continued, but by early September, fog, ice and growing darkness slowed her progress, and late in the month, only some 200 kilometres (125 miles) short of Bering Strait, Nordenskiöld was forced to establish winter quarters. He and his scientists spent the winter making detailed observations of geology, numerous studies of natural history and observations of the local Chukchi population.

On 18 July 1879, *Vega* got under way again, and two days later sailed through Bering Strait into the Pacific. Stopping along the way to build their collections and make observations about native peoples, the party steamed south to Japan, where Nordenskiöld received the first of many honours. Stops at Hong Kong, Singapore and Ceylon (Sri Lanka), and a passage through the Suez Canal preceded *Vega*'s return to Europe, where the party was feted at Naples, Lisbon, London and Copenhagen before reaching Stockholm in late April 1880.

Despite the voyage's success, the Northeast Passage still did not prove particularly viable. It was not until well into the twentieth century that the Soviet regime began to develop what has since been called the Northern Sea Route.

NORDENSKIÖLD'S EARLIER EXPEDITIONS

Adolf Erik Nordenskiöld had participated in two strictly scientific expeditions under the famed Swedish geologist Otto Torell before leading the Swedish Academy of Sciences expedition to Svalbard in 1864. Four years later, funding for science seemed to be drying up, but Dickson unexpectedly contributed the financial support for another expedition to Svalbard. On this, Nordenskiöld received international attention, because his ship *Sofia* reached a new farthest north (for a ship) of 81°42′ N. He made two more expeditions (1870 to Greenland and 1872–1873 to Svalbard) emphasizing both science and exploration towards the North Pole before adding a commercial component to his ventures

JOSEPH WIGGINS

In 1874 Joseph Wiggins revived interest in the commercial viability of the Northeast Passage by becoming the first man since the sixteenth century to reach the Ob' from Western Europe via the Kara Sea. In the following three summers, Wiggins was involved in similar efforts, and in 1878 he carried the first Siberian cargo ever to be shipped straight to England. However, unsuccessful trading ventures by others dampened interest in the route, and it was not for nine years that Wiggins made another voyage to the region. But starting again in 1887, he made seven more trading voyages across the Kara Sea.

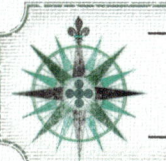

ARCTIC 1879–1882

AN AMERICAN TRAGEDY

AFTER THE AUSTRO-HUNGARIAN EXPLORING EXPEDITION'S DISCOVERIES REFUTED AUGUST PETERMANN'S THEORIES ABOUT A PATH TO THE OPEN POLAR SEA BETWEEN SVALBARD AND NOVAYA ZEMLYA (SEE PAGES 10–11), PETERMANN REVISED HIS HYPOTHESIS. INCORPORATING THE PROJECTIONS OF SEVERAL AMERICAN OCEANOGRAPHERS, HE DECIDED THAT THE BEST WAY TO REACH THE NORTH POLE WAS THROUGH BERING STRAIT, FOLLOWING THE THERMAL CURRENT KNOWN AS THE KURO SIWO TO WHERE IT JOINED THE GULF STREAM TO CUT A HOLE THROUGH THE RING OF ICE SURROUNDING THE OPEN POLAR SEA.

This notion appealed to James Gordon Bennett Jr – the proprietor of *The New York Herald*, the largest newspaper in America – who perceived a huge exclusive story in the making. Bennett therefore agreed to finance an attempt on the North Pole, although it would officially be under the auspices of the US Navy. The expedition sailed from San Francisco in July 1879 in USS *Jeannette*, under the command of George Washington De Long.

Jeannette passed easily through Bering Strait, but within a week was beset in the ice. For the next year and a half she drifted erratically, but generally northwest. In late October 1879, the party sighted Wrangel Island, which De Long correctly concluded was an island, not a continuation of a continent stretching from the North Pole, as Petermann had predicted.

The men faced a brutally hard winter, made worse when the ship was damaged by ice and the sea poured in. A pump had to be worked 24 hours a day, even as the temperature dropped to -43°C (-45°F). Spring finally came, but the ship was not released from the ice and continued to drift. In November 1880, *Jeannette* had returned to the same location she had occupied in April. Meanwhile, tensions mounted, as De Long proved to be a martinet and the two civilians on board – naturalist Raymond Newcomb and *Herald* reporter Jerome Collins – chafed under naval discipline.

In May 1881, after another winter, two of the New Siberian Islands were discovered. But on 12 June, *Jeannette* was crushed by the ice and sank. The men dragged sledges and boats to the ice edge,

ABOVE A photo of De Long taken shortly before *Jeannette* sailed. His first Arctic experience was as an officer in the search for Hall's *Polaris*.

BELOW *Jeannette* sinks after being crushed by the ice. Named for Bennett's sister, she had previously gained fame as the Arctic ship *Pandora*.

ABOVE The crew dragging the boats from *Jeannette* south across the ice. It took almost two months before they could launch them in open water.

where they launched three boats, heading for Siberia. On the way, the boats were separated in a storm, and one was never seen again.

De Long's cutter, with 14 men, landed in the north of the Lena Delta. However, the delta is a vast estuarial maze, a huge cluster of bog-like islands interlaced with countless twisting channels. After three weeks of slow progress south, one man died, and De Long ordered William Nindemann and Louis Noros to march ahead in search of relief. They eventually met with help, but it was too late for De Long and his party, all of whom died of starvation, exposure and exhaustion.

Meanwhile, the third boat, under Engineer George Melville, landed in the east of the delta, where the men met with hunters and were guided to a village where they were eventually joined by Nindemann and Noros. Melville began a search for De Long, but could find no trace of him, so, with winter approaching, he took his party up river to Yakutsk. The next spring he made a systematic search of the Delta until he found De Long's final camp.

Concurrently, a disastrous sequel was played out. In June 1881, a relief ship, USS *Rodgers*, was dispatched with William Henry Gilder, a famed *Herald* reporter, aboard. When *Rodgers* caught fire and burned, Gilder was asked to make a solo sledge trip thousands of kilometres across Siberia to Irkutsk to telegraph home news of the ship's loss. On the way, he by chance met a cossack courier to whom Melville had entrusted confidential documents about the end of *Jeannette* and the search. Gilder used these for his reports to *The Herald*. At the same time, another *Herald* reporter, John Jackson, was sent from St Petersburg to find De Long. He was eventually guided to where Melville had buried De Long and his comrades, and there he opened the tomb and searched the reporter Collins' body for correspondence, while a colleague sketched the scene. The accounts of Gilder and Jackson sent back to *The Herald* made this one of the most sensationally covered stories ever about exploration.

ABOVE Nindemann and Noros were sent ahead by De Long to bring back help for the others. They did not return in time to save their comrades.

BELOW Gilder travelled much of the world, including making an attempt on the North Pole and visiting Borneo, China and Vietnam.

GORDON BENNETT!

The *Jeannette* expedition was just one of numerous exploring efforts sponsored by *The New York Herald* and its wild and eccentric proprietor, James Gordon Bennett Jr. Through virtually unrestrained use of sensationalism, Bennett's father had built *The Herald* into the most widely distributed paper in the US. Bennett Jr continued to increase its circulation by not just reporting, but creating, the news. One of his methods was to build interest about specific locations or occurrences and then despatch an expedition to provide exclusive reports about them. His greatest success was sending reporter Henry Morton Stanley to find medical missionary David Livingstone in central Africa.

GILDER'S FIRST ARCTIC EXPEDITION

That Gilder was chosen to make the sledge trip across Siberia was not surprising with his background. In 1878, Bennett placed him as second-in-command of an American Geographical Society expedition to seek journals from the Franklin expedition that were rumoured to be still left on King William Island. Led by former US Army officer Frederick Schwatka, the party sledged from Marble Island in Hudson Bay to King William Island and back. The journey of 5,231 kilometres (3,251 miles) was the longest sledging distance then recorded, and received great publicity in Gilder's reports in *The Herald*. They found no trace of the journals from the Franklin expedition, however.

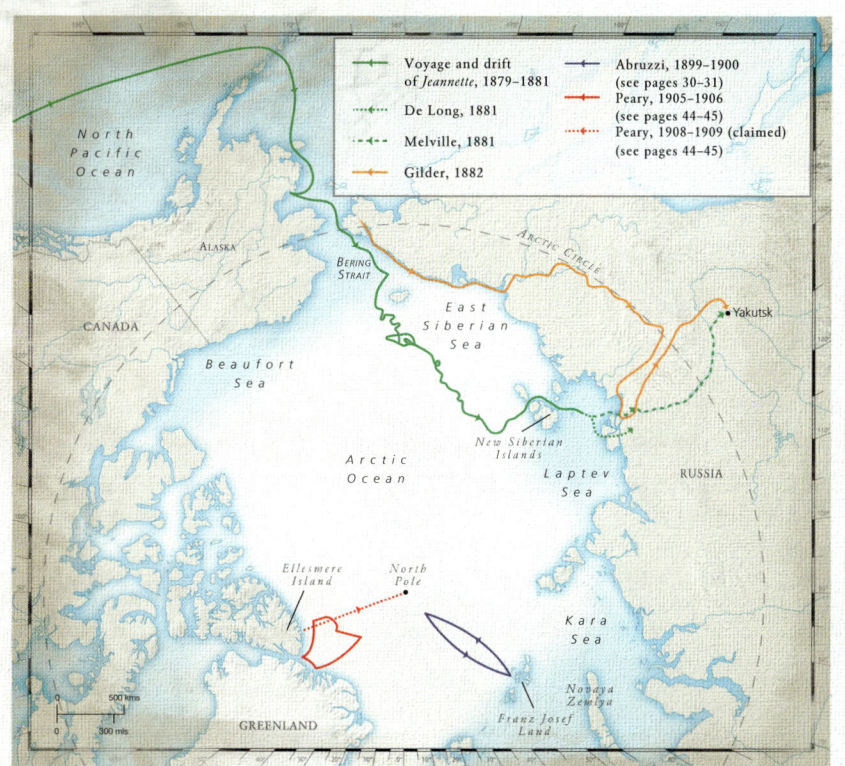

ARCTIC 1881–1884

DINING ON THE DEAD

In 1881, the US government sent a party of 25 under Lieutenant Adolphus W Greely to establish a base, named Fort Conger, at Lady Franklin Bay, on northern Ellesmere Island. Although part of the International Polar Year, it was obvious from the start that science was not the expedition's primary purpose, as none of the men had expert scientific training. Even the scientific leader and surgeon, Octave Pavy, was more interested in attaining a farthest north. Greely had similar ambitions, which were achieved in May 1882, when Lieutenant James Lockwood, Sergeant David Brainard and Fred Christiansen reached 83°24' N, besting Albert Hastings Markham's 1876 record on the British Arctic Expedition (see pages 12–13). But that was where the success stopped and the problems began.

BELOW Following the expedition Adolphus W Greely went on to a long and illustrious career. He rose to the rank of major general, served as head of the US Signal Corps, and oversaw the relief operations after the 1906 San Francisco earthquake.

That summer, the supply ship – bringing provisions, news and replacement personnel – was stopped by heavy ice and had to return south. Greely's party, which had already broken into hostile, jealous cliques, had to spend a second gloomy winter cut off from civilization.

In 1883, sledge parties charted the interior of Ellesmere Island and discovered the vast Agassiz Glacier, but once again the supply ship did not arrive. So, in August, the men set off south in two open boats and a steam launch. They moved slowly through the pack ice between Ellesmere and Greenland and finally arrived in late September at Cape Sabine on Pim Island, where supplies were to have been left by two relief ships.

Unfortunately, when the primary relief ship had arrived at Cape Sabine with a substantial cache several months before, the inexperienced commanding officer had seen an open lead toward the north, so had steamed off without unloading any materials. Within hours, his ship was crushed in the ice and sank, forcing the crew to make its way south in open boats. Learning of this loss, the commander of the second ship turned south after them, also not depositing any supplies.

Thus, when Greely arrived, without the provisions to continue south, his party was forced to winter there. They built a metre- (three-foot-) high stone hut with the whaleboat as the roof, an accommodation that was far too small for 25 men. With the food almost gone, Greely soon cut the rations to 280 grams (10 ounces) per man per day. In the brutally cold winter that followed, Joseph Elison was so badly frostbitten that his feet and hands had

ABOVE An artist's conception of the inside of the final tent at Cape Sabine before the rescue. In reality, the tent had partially collapsed, and the men inside did not have nearly so much space.

ABOVE Adolphus W Greely (third from right), standing at the front next to Commander Winfield Scott Schley, who commanded the successful rescue expedition. Surrounding them are other members of the rescue operations.

to be amputated, and then in January, Sergeant William Cross died of starvation and scurvy. A similar fate for the others was postponed by the bagging of a small bear, a remarkable feat in an area virtually devoid of wildlife.

In March 1884, with the return of the sun, Sergeant George Rice started fishing for shrimp, but these were so tiny that 700 were needed to produce just 28 grams (an ounce) of meat. Men continued to die, existing only on the shrimp, bits of the plant saxifrage and their sealskin boots and coats. In April, five men, including Lockwood and Rice, died within a week. The others were sustained only by the capture of a seal. Brainard, who took over the shrimping, was the only man able to work regularly until Pavy suddenly began to disappear daily, claiming he was chopping ice to melt for water.

In May, when melt water made the hut uninhabitable, the 14 survivors moved into a tent. At the beginning of June, Charles Henry, who had been repeatedly caught stealing food, was executed. Pavy died shortly thereafter, apparently from an unintentional overdose of ergot (a fungus that infects cereal plants). Then, on 22 June, the horror ended. The seven remaining men were rescued by a relief expedition. When Greely was found under the collapsed tent, he could only gasp: "Here we are – dying – like men. Did what I came to do – beat the record." Although Elison died shortly thereafter, the other six, including Greely and Brainard, returned home as heroes.

But the celebration did not last long, as tales of cannibalism soon surfaced, and it turned out that Pavy had been eating his dead comrades during his frequent disappearances in the weeks before his death. The horrible, graphic accounts subsequently splashed on the front pages of newspapers around the world gave the expedition sensational press coverage. ✤

- Markham's farthest north, 1876 (see page 12)
- Lockwood's farthest north, 1882
- Greely's retreat to Cape Sabine, 1883

RIGHT Joseph Pulitzer, who turned the *St Louis Post-Dispatch* into a journalistic power before using sensational techniques to help build the largest daily circulation in the US at *The World* of New York.

THE INTERNATIONAL POLAR YEAR

After the return of the Austro-Hungarian Exploring Expedition in 1874, Karl Weyprecht, its joint leader, publicly stated that the era of expeditions aimed at geographical exploration to the exclusion of science was over. He proposed that, rather than "a sort of international steeplechase … primarily to confer honour on this flag or the other," a co-ordinated, international programme of scientific investigation be established. The ultimate realization of his ideas was the International Polar Year, 1882–1883, in which 14 scientific stations were supposed to make simultaneous scientific observations in order to discover fundamental laws and principles of nature.

LEFT Karl Weyprecht, co-leader of the Austro-Hungarian Exploring Expedition. Weyprecht's concepts were developed into the International Polar Year.

THE SENSATIONAL COVERAGE

When word was received of Greely's rescue, the entire front page (or more) of virtually every major American newspaper was given over to the story. During the ensuing weeks, prominent coverage was given to the homecomings or funerals of expedition members. But nothing could equal the furore launched on 12 August, when *The New York Times* first drew attention to Henry's execution and the charges of cannibalism. Perhaps the most sensational account of cannibalism was plastered across the front page of the largest newspaper in the country, Joseph Pulitzer's *The World*.

ARCTIC 1888–1889
NANSEN CROSSES GREENLAND

One of the great mysteries of the Arctic throughout most of the nineteenth century was what might lie in the immense, unknown interior of Greenland, beyond the ice that seemed to continue without pause from near the coasts. Adolf Erik Nordenskiöld, the conqueror of the Northeast Passage and one of the most respected Arctic scientists in the world, had long believed that in the centre of the vast island, high up the ice cap, was a fertile oasis. In 1883, he led an expedition that was the first to penetrate any great distance on to the ice cap. However, in doing so, it was determined that, in all probability, the ice fully covered the island's centre. Three years later, the American explorer Robert E Peary set out to become the first to cross Greenland, but he was forced by conditions and lack of supplies to turn back 160 kilometres (100 miles) inland.

In late 1887, a young Norwegian scholar, Fridtjof Nansen, stepped on to the scene when he formally announced his intention of making the first crossing of the ice cap. Nansen's plan was innovative to the point of being audacious. First, he would travel on skis, which had never been tested at the altitude and under the conditions of the ice cap. Instead of starting from the inhabited west coast, as had previous expeditions, his party would be dropped near the sparsely populated east coast of Greenland. This not only meant that Nansen had to cross the ice only once, but that there could be no hesitation and no turning back: salvation was only available in the form of supplies left on the west coast. Nansen received little support, either financially or in the press, but his expedition

ABOVE Fridtjof Nansen around the time of the first crossing of Greenland. It has been argued that Nansen had a more profound combination of brilliance, insight and originality than any other explorer in history.

RIGHT A drawing of Nansen (centre) and the five men who accompanied him on his Greenland expedition. In font of them are a sledge and examples of gear they used in crossing the island's ice cap.

LEFT A bust of Fridtjof Nansen, held at the Royal Geographical Society. As well as being one of the greatest explorers, Nansen served as professor of both oceanography and zoology, was a key diplomat for the young Norwegian nation and earned the Nobel Peace Prize.

they made their way back north, to a new starting point near Umivik. On 15 August, they began their inland journey, heading northwest towards Christianshåb on Disko Bay, at about 69° N. But warm temperatures and then gales slowed their progress on to the ice cap, forcing Nansen to change their destination from Christianshåb to Godthåb, which was much closer, at only 64° N.

On 2 September, the party was able to start using skis, and the journey continued more or less uneventfully, proving how well adapted the mode of transport was to polar travel. Although skis had been used in different environments before, Nansen improved traditional techniques by using waxless telemark skis, innovative steel edges and bindings with a heel grip rather than just a toehold.

The party passed the highest point of land on 12 September, and a week later they sighted the west coast. Sails were raised over their sledges as they made their descent down the western side of the ice cap. They reached the sea on 26 September, having descended through the valley of Austmannadalen to the head of a fjord. Over the next three days, they built a boat, which Nansen and Otto Sverdrup rowed to Godthåb. There they were greeted by the local Danish governor, who congratulated Nansen not only on the crossing, but also for having had his doctorate conferred upon him in his absence.

The last ship of the year to Europe had already left, so the party wintered at Godthåb, where Nansen further developed sledging equipment, the technology of skis and the techniques for driving dogs. By the time the party reached Norway in May 1889 – having shown with certainty that the ice cap extends unbroken across Greenland – Nansen had become an international hero. ✣

NORDENSKIÖLD ON THE GREENLAND ICE CAP

In 1883, Nordenskiöld travelled to Greenland to determine whether the ice cap covered the whole of the island's interior. His party moved inland in July, covering 117 km (72 miles) and reaching an altitude of 1,510 metres (4,953 feet) in 17 days. Here, Nordenskiöld decided to turn back, but he sent two Lapp ski experts to assess the conditions farther east. They returned 57 hours later, claiming to have travelled 230 km (143 miles) each way without finding land. Although later explorers suggested that distance was exaggerated, it has since been confirmed that the ice cap continues over most of Greenland.

ABOVE Nordenskiöld's assault on the Greenland ice cap was unsuccessful.

NANSEN'S TECHNOLOGICAL INNOVATIONS

Although showing the application of skis to polar travel was Nansen's most obvious contribution while crossing Greenland, there were other areas in which he broke new ground, too. Most notably, he replaced the traditional, narrow-runnered sledge with a prototype of the modern one – lighter, flexible, and running on skis, one known today as the Nansen sledge. He also designed special clothing, tents, and cooking equipment, including the "Nansen cooker", a saucepan that conserves heat and fuel. And he was a pioneer in applying scientific principles to the calculation of the dietary needs of expedition members.

was able to proceed after the Danish businessman Augustin Gamél volunteered to fund it.

On 17 July 1888, Nansen and five others left the ship *Jason* near Sermilik Fjord, where they hoped to start their trek. But their boats became trapped in the ice and for almost two weeks they drifted farther and farther south, leaving behind their intended landing location. Finally, at the end of the month, they managed to row ashore, and from there

BELOW Nansen with a sledge and a dog. Perhaps the most significant of Nansen's technological developments was the versatile Nansen sledge. He also was a pioneer in understanding the use of dogs in long-distance polar exploration.

ARCTIC 1893–1896

THE VOYAGE OF *FRAM*

IN 1884, RELICS FROM *JEANNETTE* (SEE PAGES 16–17) WERE DISCOVERED NEAR JULIANEHÅB, IN SOUTHWEST GREENLAND. SOME FIVE YEARS LATER, AFTER HIS RETURN FROM CROSSING THE ISLAND, FRIDTJOF NANSEN CONJECTURED THAT THE SAME CURRENT THAT CARRIED THOSE ITEMS OVER THE TOP OF THE WORLD MIGHT BE USED TO EXPLORE THE WATERS OF THE ARCTIC BY SHIP. IN FACT, HE THEORIZED, IF A SHIP WERE INSERTED INTO THE ICE PACK NORTH OF SIBERIA AT THE CORRECT PLACE, THE DRIFT MIGHT TAKE IT DIRECTLY ACROSS THE NORTH POLE BEFORE RELEASING IT ON THE OTHER SIDE OF THE ARCTIC OCEAN.

One of Nansen's first steps in proceeding with his idea was to have a specially designed ship constructed that would not be crushed in the ice. Designed by Colin Archer, she was named *Fram* ("Forward"), and her sides sloped sufficiently to prevent the ice from getting a firm hold on her hull; when squeezed between ice floes, she simply rose up onto the ice. In addition, *Fram* was equipped with only the third marine diesel engine ever installed in a ship; she was furnished with electric lights, the dynamo for which could be driven by the engine, the wind, or hand-power; and she incorporated numerous features to make her a floating scientific research station.

Despite widespread scepticism about Nansen's proposal, in 1893 *Fram* left Norway under the command of Otto Sverdrup, one of Nansen's comrades in the crossing of Greenland, who would also serve as second-in-command of the expedition. After passing along most of the Northeast Passage, Nansen intentionally allowed *Fram* to be frozen into the ice near the New Siberian Islands, and he, his small party and 34 dogs disappeared into the mysterious north. In the following year-and-a-half, Nansen oversaw a wide-ranging scientific programme, and proved that the Eurasian side of the Arctic Basin was a true, deep ocean.

However, as time passed, it became apparent that the drift would not take *Fram* over the North Pole as Nansen had hoped, so he decided to make a dash for it. In March 1895, he and former gymnast Hjalmar Johansen left the ship to try to reach the Pole with 27 dogs, three sledges and two kayaks, despite knowing that, in an ever-moving sea of ice, they would never be able to find the ship again. For almost a month they battled on, until, having reached 86°13′ N – 313 kilometres (169 geographical miles) closer to the North Pole than anyone had previously been – Nansen determined that the southward drift of the ice that they were then on would not allow them to reach the Pole.

Nansen and Johansen turned towards little-known Franz Josef Land, but made slow progress owing to hummocky ice conditions and wide lanes of open water. In August, they finally reached the archipelago and began making their way south, although, since the chronometers they carried had at one point stopped, they were unable to measure longitude precisely and were uncertain if they were actually in Franz Josef Land or another undiscovered set of islands. Conditions became so difficult in September that they were forced to halt and spend a long, miserable winter in a makeshift hut living on polar bear and walrus

TOP *Fram* trapped in the ice. Although dreaded by most ships, this is what she had been designed for, and it was more comfortable to live in her in the ice than in open water.

ABOVE In March 1895, after her second winter in the Arctic, the crew of *Fram* dug her out of the ice that had built up around – and over – her.

ABOVE Hjalmar Johansen at Jackson's hut at Cape Flora. Nansen was so formal that although he and Johansen shared a sleeping bag on their northern dash, he never allowed his subordinate to address him in the informal mode of speech.

BELOW Nansen on the ice near *Fram* taking a deep-water temperature as part of the expedition's extensive oceanographic studies. A glass lantern slide of this picture was used as part of Nansen's later lectures.

ENCLOSURE

1. A letter from Fridtjof Nansen to Sir Mountstuart Grant Duff, the president of the RGS, asking for his help in obtaining a military balloon for research during the upcoming Arctic drift.

2. The plan of *Fram*, after her refitting for Otto Sverdrup's expedition (1898–1902). Sverdrup being captain on Nansen's expedition gave him clear insight into the needed improvements. (See pages 26–27.)

BELOW Two members of Nansen's expedition on *Fram* take measurements to help determine their exact location. The man in the front is using the theodolite, while the man behind studies a chronometer.

ABOVE Benjamin Leigh Smith, whose scientific studies earned him the Patron's Medal of the Royal Geographical Society.

meat. Finally, in May 1896, they continued south through the archipelago, bearing, they hoped, for Cape Flora, where Benjamin Leigh Smith's old base had been. The next month, they had a remarkable – and extremely fortunate – meeting with Frederick Jackson, an English explorer whose expedition had made its base just where the Norwegians were heading. After enjoying Jackson's hospitality for a period of weeks, Nansen and Johansen returned to Tromsø on his ship.

Meanwhile, *Fram*, under Sverdrup, had continued her drift, attaining a high latitude of 85°55′N, and then finally reaching land at Danskøya, off northwest Spitsbergen, where Salomon Andrée's balloon expedition attempting to reach the North Pole was encamped (see pages 24–25). Sailing south shortly thereafter, *Fram* reached Tromsø just one day before Nansen, who had been proven correct in every respect and whose expedition had produced incalculably valuable scientific results.

BENJAMIN LEIGH SMITH IN FRANZ JOSEF LAND

Nansen was not the first explorer forced to winter in Franz Josef Land. Benjamin Leigh Smith, a gentleman-scientist, had made three summer voyages to Svalbard before exploring the western islands of Franz Josef Land in 1880. He returned the next year to extend his survey, but while he was collecting fossils and plants at Cape Flora, his ship, *Eira*, was crushed in the ice. The crew retrieved many stores before she sank, and, after a relatively comfortable winter, they took to open boats and sailed and rowed to Novaya Zemlya, where they met a rescue party.

THE JACKSON-HARMSWORTH EXPEDITION

In 1894, with the financial backing of Alfred Harmsworth, Frederick Jackson launched an expedition to explore Franz Josef Land and, he hoped, make an attempt on the North Pole. Harmsworth (later Lord Northcliffe) – who had made a fortune with his weekly magazine *Answers* – had just purchased his first newspaper, *The Evening News*, and saw the expedition as a chance to gain readership with exciting, exclusive reports. However, although Jackson did a superb job mapping the archipelago, he made no serious attempt on the Pole, and the expedition is remembered chiefly for his encounter with Nansen.

ABOVE A staged photograph to recreate the meeting of Nansen (right) and Frederick Jackson near Cape Flora.

ARCTIC 1896–1897: BALLOON TOWARDS THE POLE

When, shortly after concluding her drift, *Fram* reached the tiny island of Danskøya, off northwest Spitsbergen, Otto Sverdrup found an amazing thing – a party hoping to launch a hydrogen-filled balloon (or aerostat) as a means of reaching the North Pole. The expedition was led by the Swedish engineer Salomon August Andrée, who had stunned the world's scientific and geographical communities the previous year when he revealed his audacious plan.

Andreé projected his balloon flight would be about 4,000 kilometres (2,600 miles), more than twice what had previously been accomplished anywhere on Earth, and would take a duration of six days, although the longest previous flight had been only 24 hours. A number of scientists had been sceptical, not only about his being able to meet these projections, but also about his ability to prevent the loss of hydrogen from the balloon envelope and to deviate in his course from the direction of the wind. Nevertheless, Andreé was so persuasive that he received financial backing from the wealthy industrialist Alfred Nobel, Nordenskiöld's former sponsor Oscar Dickson and King Oscar of Sweden.

In June 1896, Andrée, his two balloon companions – Nils Eckholm and Nils Strindberg – and a team of scientists and technicians sailed to Danskøya. There, a hangar was built, a generator installed, gas-making equipment set up and the French-designed *Örnen* (*"Eagle"*) inflated. But the wind started blowing from the north – the opposite of what Andrée wanted – and it continued unabated for three weeks. In mid-August, Andrée gave up in despair and returned to Europe.

The next year, he was back to try again, accompanied by Strindberg and Knut Frænkel, who had replaced Eckholm when the latter resigned because of concerns about safety. On 11 July, *Örnen* was finally launched, but ropes trailing from the gondola that housed the men immediately snagged on a large rock, and the entire balloon was nearly pulled down into the sea. Andreé righted the envelope, and the three floated away into the mists to the north. No one ashore realized that the accident had ripped away three trailing ropes designed to control direction and maintain altitude. Furthermore, to prevent *Örnen* crashing into the sea, 675 kilograms (1,485 pounds) of ballast had been thrown overboard, so the balloon rose to a far greater height than intended, leading to a loss of gas. These developments meant that before *Örnen* ever disappeared from sight, her manoeuvrability and aerial longevity had been severely compromised.

In fact, the men had basically lost any real control of the balloon, which drifted aimlessly, at points bumping on the sea ice, before, two days later, being brought to rest on the Arctic pack ice some 350 km (220 miles) northeast of Svalbard. Removing sledges, food and equipment from the gondola, they started southeast on foot towards Franz Josef Land, engaging in an arduous struggle over the ice. However, the drift of the ice was carrying them directly south – which meant they would miss Franz Josef Land – so they changed direction, hoping to reach Sjuøyane in northern Svalbard. But the ice again frustrated them, and in September, they saw not Sjuøyane, but the previously unexplored island Kvitøya in the far northeast of Svalbard. On 5 October,

TOP LEFT Andrée and his party drift northward in *Örnen* over Virgo Harbour as they depart Danskøya. The men on the shore had no idea that none of the three aeronauts would ever be seen alive again.

LEFT Members of Andrée's expedition testing the balloon envelope of *Örnen* for leaks. She was pronounced in perfect condition, but lost much hydrogen after taking off when she rose to a greater height than intended.

BELOW A French artist's conception of the beginning moments of Andrée's flight from Danskøya. In fact, the gondola was larger than shown here.

they made the first-ever landing on the island, where they set up camp. Soon thereafter, their diaries ended and it is likely they died within weeks, perhaps from carbon monoxide poisoning from a stove, from cold and exposure, or from trichinosis, caused by eating under-cooked polar bear meat.

In the ensuing years, numerous searches were made, but the party's fate remained a mystery. For more than 30 years the world could only wonder what had become of them. Then, in August 1930, a Norwegian fishing vessel landed at Kvitøya and discovered the camp. Diaries, logbooks, canisters of film – still holding usable pictures – and other relics were located, telling the entire disastrous story. The men's remains were taken back to Sweden, where they were buried with honour in Stockholm's cathedral.

RIGHT Although remembered primarily as a mountaineer, Sir Martin Conway was also Professor of Fine Arts at Cambridge and a Unionist Member of Parliament.

LEFT Two of the aeronauts stand near *Örnen* after her landing on the sea ice. The film was recovered with the men's bodies more than three decades later.

THE FIRST CROSSING OF SPITSBERGEN

The same two summers that Andrée was on Danskøya, a British explorer, Sir Martin Conway, was conducting the first extensive exploration of Spitsbergen's interior. In 1896, sponsored by the Royal Geographical Society, Conway, the respected geologist J W Gregory and E J Garwood became the first men to cross Spitsbergen. They used ponies to pull their sledges while crossing from Adventfjorden (near where Longyearbyen was later founded) to Agardhbukta on the east coast and back again. In 1897, Conway and Garwood returned and explored the interior, north of where they had been the previous year.

WALTER WELLMAN'S AIRSHIP FLIGHTS

The decade after the disappearance of Andrée, Danskøya again became the scene of aerial efforts to reach the Pole. In 1906, Walter Wellman – who had previously led unsuccessful sledging attempts on the North Pole from Svalbard and Franz Josef Land – set up near Andrée's base, intending to reach the Pole in a hydrogen-filled, powered airship (dirigible). Serious defects in the equipment prevented a departure, but Wellman returned in 1907, only to have contrary winds force the airship down after four hours. Wellman made a final attempt in 1909, but after going only 50 kilometres (32 miles), an accident made the airship unmanageable.

BELOW Walter Wellman on the wing of his airship. Full of thrilling ideas, Wellman never was able to turn them into success.

BELGICA'S ANTARCTIC WINTER
Antarctic 1897–1899

In the 1890s, the geography of Antarctica was still a total mystery. Was it a continent? Was it a set of ice-covered islands like Franz Josef Land? No one knew with certainty, but the scientific, geographical and commercial communities increasingly supported its investigation. Then, in 1895, the Sixth International Geographical Congress was held in London. At the urging of Sir Clements Markham of the Royal Geographical Society, the conference passed a resolution stating that the Antarctic was the most significant geographical area remaining to be investigated and that it should be explored before the turn of the century.

ABOVE In the years after the expedition, Émil Racovitza gave up his zoological studies of the oceans and began research into cave fauna. He later founded the world's first speleological institute.

In fact, preparations were already underway. An expedition left Antwerp in 1897, and, although being primarily Belgian financed, sailing under the Belgian flag, having a Belgian commander – Adrien de Gerlache – and heading south on a former whaler renamed *Belgica*, it was, in fact, the first truly international Antarctic expedition. Geologist Henryk Arctowski and assistant geologist Antoni Dobrowolski were Polish. Zoologist Emil Racovitza was Romanian. Much of the crew was Norwegian, including the 25-year-old mate, Roald Amundsen. The ship's surgeon was Frederick A Cook, an American who had served under Robert E Peary in the Arctic.

De Gerlache spent an unexpectedly long period conducting research in Tierra del Fuego, and thus *Belgica* did not arrive in Antarctic waters until 20 January 1898, late in the exploring season. Two days later the first tragedy occurred when Karl Wiencke, a sailor, was washed overboard in a gale and drowned. During the following three weeks, *Belgica* zig-zagged down Gerlache Strait between a series of islands and the Antarctic Peninsula. They made about 20 landings, conducting investigations and naming numerous islands and other features.

In mid-February, *Belgica* crossed the Antarctic Circle, and two weeks later she entered the ice pack. By early March she was imprisoned in the ice, and began drifting southwest at a rate of some 8–16 kilometres (5–10 miles) a day. Realizing that they could not now escape, many of the men became irritable and despondent, and their condition worsened after 17 May, when the sun set and they began the first wintering south of the Antarctic Circle. The situation became worse when, at the beginning of June, the geophysicist, Lieutenant Danco, died from a heart condition. It was perpetually damp and cold, and, because of the inadequate vitamins provided by the canned food, some men began to suffer from scurvy.

At this point, de Gerlache seemed almost to withdraw from command, concentrating primarily on taking care of the ship. But with the party falling apart, Cook effectively took command, using the knowledge he had gained in the Arctic. He forced the men to eat seal and penguin to ward off scurvy, and he kept open fires burning several hours per day, hoping that the light and warmth would improve the crew's morale and physical condition. Despite this, a number of the crew showed signs of madness, and it was not until 21 July,

LEFT *Belgica* trapped in the ice. The crew's collapse of morale showed de Gerlache did not understand the lessons British officers had learned while in similar conditions in the high Arctic (see pages 12–13).

TOP Adrien de Gerlache formulated the idea for a Belgian Antarctic expedition while serving on a ship making hydrographical studies. In later years, he helped Jean-Baptiste Charcot organise his first expedition.

ABOVE Although Henryk Arctowski was unsuccessful in organising a second Belgian Antarctic expedition, he remained an important international figure in geophysics and meteorology for more than half a century.

ABOVE LEFT After serving on *Belgica*, Roald Amundsen became the most accomplished polar explorer ever. He was the first to navigate the Northwest Passage, first to attain the South Pole and first to reach the North Pole.

when the first edge of the sun appeared in the horizon that they began to pick up.

But *Belgica* remained trapped in the ice, and conditions did not improve as summer approached, leaving the expedition members terrified that they might have to spend another year there. On the last day of 1898, a stretch of open water appeared only 640 metres (700 yards) away. At this point, Cook and Racovitza suggested blasting the ice away between the ship and the water and cutting a channel to it. Daily progress was made, and by the end of January they were only 30 metres (33 yards) away from open water. But their hopes were crushed when, owing to a change of wind, the ice field shifted and the sides of the channel squeezed together.

Two weeks later, things changed equally suddenly for the better, as the channel opened again. The engines were started, and for the first time in a year *Belgica* moved on her own power. But there were still 11 kilometres (seven miles) of pack ice between them and the open sea, and the next month was a constant struggle, before *Belgica* finally emerged free from the ice on 14 March 1899.

RIGHT Henrik Bull, leader of the expedition sent by Svend Foyn in *Antarctic*. They took elephant seals at Iles Kerguelen, but were unsuccessful in finding many whales.

BELOW Members of the crew of *Belgica* taking soundings through the ice. As well as the scientific data, one achievement of the expedition was taking the earliest known photographs of the Antarctic proper.

SEARCHING FOR NEW WHALING GROUNDS

The late nineteenth century saw a decline in the whaling industry. Between 1892 and 1895, four expeditions went to investigate whether enough baleen whales could be found in the Antarctic to make a southern industry economically viable. Two were Norwegian expeditions led by Carl Larsen, who would later found the South Georgia whaling industry. Another Norwegian expedition, under Henrik J Bull, made the first continental landing from the region of the Ross Sea (earlier ones were on the peninsula). A Scottish expedition proved to several scientists who accompanied it – including W S Bruce – that there were vast scientific gains to be made from Antarctic research.

THE SIXTH INTERNATIONAL GEOGRAPHICAL CONGRESS

The resolution passed at the Sixth International Geographical Congress was a key development in re-starting Antarctic exploration. It stated: "That this congress record its opinion that the exploration of the Antarctic Regions is the greatest piece of geographical exploration still to be undertaken. That, in view of the additions to knowledge in almost every branch of science which would result from such a scientific exploration, the Congress recommends that the scientific societies throughout the world should urge, in whatever way seems to them most effective, that this work should be undertaken before the close of the century."

ANTARCTIC 1898–1900: WINTERING ON THE CONTINENT

Carsten Borchgrevink, a Norwegian who had resettled in Australia, came to public attention after claiming to have been the first person to stand on the Antarctic continent. At the Sixth International Geographical Congress in London (see pages 26–27), he put himself forward to lead an expedition that would winter on the continent, but failed to gain support. Further, he was actively opposed by the president of the Royal Geographical Society, Sir Clements Markham, who was raising funds for a British national expedition (see pages 32–33). Forced to turn elsewhere, Borchgrevink approached publishing magnate George Newnes about funding his Antarctic venture, and Newnes responded with a grant of £40,000.

At the insistence of Newnes, the expedition sailed under the British flag, but most of its members were Norwegian, with two Lapps and a sprinkling of others. The ship *Southern Cross* sailed from London in August 1898, and left Hobart, Tasmania, for the Antarctic in December. They spent weeks getting through the pack ice of the Ross Sea, and it was not until mid-February 1899 that the party reached Cape Adare at the northern tip of Victoria Land, where Borchgrevink planned to winter. Before all of the building materials and supplies were unloaded, a furious gale battered *Southern Cross*, and a party of seven was marooned ashore. The physicist Louis Bernacchi later told of how they were only prevented from freezing to death because the dogs came into the tent and lay on top of them. Eventually, two prefabricated huts were assembled – Borchgrevink naming the base Camp Ridley for his mother's maiden name – and 10 men were left there, the first ever to spend a winter on the continent.

Over the next year, despite terribly harsh conditions, extensive scientific data was obtained. But there were also near disasters, and bitter feuds between Borchgrevink, who was not a natural leader, and his scientific staff. In June, Persen Savio fell to the bottom of a crevasse – he cut toeholds with a knife in order to climb out. In July, the huts narrowly avoided being destroyed by fire after a candle set a bunk alight. In August, three men asleep in a hut were nearly asphyxiated by fumes from unattended coals in the stove; fortunately, Bernacchi managed to get the door open. Then in October, disaster could not be averted: Nikolai Hanson, the zoologist, died of undetermined causes after a lingering illness. He was buried atop a local hill in a grave that had to be blasted out of the rock with dynamite.

ABOVE The cliffs at Cape Adare. The landing site that the base was established on consisted of a small, low-lying beach that could be extremely difficult to reach because of bad weather and high winds.

RIGHT Although only in his early twenties, Louis Bernacchi was a talented scientist who conducted much of the meteorological and magnetic work and was in charge of photography on the expedition. He later joined the *Discovery* Expedition (see pages 32–33).

ABOVE Camp Ridley, with the main hut and its surroundings covered by snow from a blizzard. The hut measured 4.6 metres (15 feet) on each side.

In late January 1900, *Southern Cross* finally arrived back at Cape Adare to pick up the expedition members. They landed on several islands in the Ross Sea, including Ross Island, where Borchgrevink and Bernhard Jensen, captain of *Southern Cross*, were nearly swept out to sea by a wave caused by a large iceberg calving. Leaving Ross Island, they sailed some 600 kilometres (375 miles) along the Great Ice Barrier (Ross Ice Shelf), and found that it had receded 48 kilometres (30 miles) since James Clark Ross had first sighted it.

When the party reached the break in the Barrier that Ernest Shackleton later named the Bay of Whales, Borchgrevink, the magnetician William Colbeck and Savio sledged south for eight hours, attaining an estimated latitude of 78°50′ S, the farthest south reached up to that date. Three days later, Bernacchi and three others repeated the feat, reaching roughly the same latitude. The expedition then turned north, and after a stop at Franklin Island for magnetic observations, they headed to New Zealand.

Despite the farthest south and a number of successful scientific studies, Borchgrevink never received a great deal of recognition. When he returned to Britain, he found the press and the scientific and geographical communities focusing on the British National Antarctic Expedition (1901–1904), which Markham had finally seen carried beyond the planning stages. It was not until 1930 that the Royal Geographical Society belatedly recognized the value of Borchgrevink's pioneering work in the far south.

ABOVE Camp Ridley as it looks today. Changes were made to the base when Scott's Northern Party later wintered there, and some of the buildings have since been protected and conserved.

BELOW Borchgrevink (right) and one of his companions at an estimated latitude of 78°50′S, a new record for the farthest south on the Great Ice Barrier.

Borchgrevink's Claim

In September 1894, Borchgrevink joined Henrik J Bull's expedition searching for right whales in the Ross Sea. They failed to find many whales, but in January 1895, Bull and six others set out to land at Cape Adare. Just who actually reached the shore first was later hotly debated, but Borchgrevink reported in a magazine article that he jumped ashore before the others, making him the first man to ever step on the Antarctic continent. His story was widely accepted, and the publicity accompanying it gave him credibility when he approached Newnes for funding.

ABOVE An artist's impression of Borchgrevink's landing on the Antarctic continent, at Cape Adare.

First on the Antarctic Continent

Although Borchgrevink was long thought to be the first man to set foot on Antarctica, he was actually preceded in that feat by some 75 years. On 7 February 1821, John Davis, a sealer from New Haven, Connecticut, made the first recorded landing on the Antarctic continent, at Hughes Bay on the Peninsula. But, as with many of his contemporaries, Davis was only interested in new lands for their commercial value, so he failed to announce his discoveries. It was not until the 1950s that his logs were studied and his achievements realized.

LEFT Carsten Borchgrevink's favourite studio portrait of himself.

ARCTIC 1898–1902
SECOND *FRAM* EXPEDITION

SOON AFTER THE RETURN OF THE *FRAM* EXPEDITION, FRIDTJOF NANSEN AND OTTO SVERDRUP BEGAN PLANNING ANOTHER MAJOR EXPLORING EFFORT. THEIR IDEA WAS TO TAKE *FRAM* THROUGH SMITH SOUND AND PROCEED AS FAR AS POSSIBLE ALONG THE NORTHWEST COAST OF GREENLAND. WHEN STOPPED BY ICE, THEY WOULD PROCEED WITH DOG-SLEDGES TO FOLLOW THE COASTLINE AROUND ITS NORTHERN TIP AND THEN DOWN ITS WESTERN SIDE, ANSWERING ALL THE QUESTIONS ABOUT THE GEOGRAPHY OF THIS MYSTERIOUS REGION. THEIR PLAN RECEIVED THE BACKING OF THREE WEALTHY KRISTIANIA (OSLO) BUSINESSMEN, AXEL HEIBERG AND THE BROTHERS AMUND AND ELLEF RINGNES.

For a variety of reasons, Nansen was unable to go, so Sverdrup took command. Two months after sailing in July 1898, heavy ice halted the progress of *Fram* near Cape Sabine, where Adolphus Greely had wintered his final year (see pages 18–19). Sverdrup was thus forced to establish a camp on the nearby coast of Ellesmere Island. In October, during one of his exploratory journeys, he briefly met the American explorer Robert E Peary, whose own expedition was wintering in the area (see pafes 44–45).

The next spring, Sverdrup and Edvard Bay crossed southern Ellesmere and, from the head of Bay Fiord, became the first men ever to see Axel Heiberg Island. That July, Sverdrup attempted to take *Fram* north. However, a month trying to pass through the ice proved unsuccessful, and Sverdrup finally decided to abandon the idea of reaching the north coast of Greenland. He revamped his plans to investigate the uncharted area to the west of Ellesmere. To advance this strategy, he moved the ship to Harbour Fiord off Jones Sound, along the south coast of Ellesmere.

Throughout 1900 and 1901, parties from *Fram* explored and charted the coasts and inland parts of Ellesmere, discovering and mapping much of the area of three other large islands of the region: these were named Axel Heiberg, Ellef Ringnes and Amund Ringnes.

During the extensive sledging conducted throughout the period, Sverdrup extended and perfected the techniques of polar travel that Nansen had developed (see pages 20–21), including demonstrating that skis could be used on most kinds of snow and sea ice and improving the equipment used for sledging. But his greatest technical contributions were in initiating the creative interplay between skis and dogs, proving not only that skis could be used for men to keep up with dogs, but that a man skiing at the correct pace actually goes the same speed as a dog pulling a sledge. This meant that dogs could be driven by men skiing instead of riding on sledges, so that bigger loads could

ABOVE Nansen on the ice during the drift of *Fram*. He hoped to lead an expedition to northern Greenland, but after the planning was well under way, family commitments forced him to withdraw in favour of Sverdrup.

BELOW *Fram* in the ice. The greatest of polar ships, she was used for three expeditions: Nansen's polar drift, Sverdrup's opening up of the Canadian archipelago and Amundsen's attainment of the South Pole.

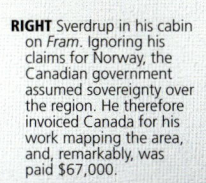

RIGHT Sverdrup in his cabin on *Fram*. Ignoring his claims for Norway, the Canadian government assumed sovereignty over the region. He therefore invoiced Canada for his work mapping the area, and, remarkably, was paid $67,000.

FAR RIGHT Peary at the time of Sverdrup's expedition. He could not conceive how one could be in the region and not have the North Pole as a goal.

Peary's Midwinter Race

Peary had already led five expeditions to north Greenland when he turned his eyes towards the North Pole. When, in 1898, Peary found that Sverdrup was also wintering on Ellesmere, he quickly became obsessed with the notion that the Norwegian was aiming for the Pole. Desperate to beat Sverdrup to Greely's old base at Fort Conger, which could be used as a launching point for the Pole, in December Peary made a midwinter dash there. During the 18 days it took to reach Fort Conger, Peary's feet were severely frostbitten, and ultimately most of his toes were amputated. After spending much time recovering, it was not until 1902 that Peary made his first concerted effort to reach the Pole, reaching only 84°17′ N (see page 44).

A Farthest North

While Sverdrup and Peary were on Ellesmere, the famous climber Luigi Amedeo di Savoia, the Duke of Abruzzi, led another attempt on the North Pole. In 1899, his Italian expedition sailed all the way to Rudolf Island, in northern Franz Josef Land. That winter, Abruzzi's fingers were severely frostbitten, forcing him to turn over field command to Umberto Cagni, who set out in March 1900. One of his support parties vanished, but Cagni nevertheless continued until a lack of provisions forced a return. He eventually reached 86°34′ N, breaking Nansen's farthest north (see page 22) by 39 kilometres (21 geographical miles).

RIGHT A world-class mountaineer, the Duke made first ascents in the Alps, Alaska and central Africa, and reached a record altitude in the Karakoram.

be carried and greater distances attained. Sverdrup also showed that Europeans could drive dogs as effectively as the Inuit, and that dogs from Greenland were more able to stand the constant rigours of long Arctic field trips than the Siberian breeds that had previously been popular with many Europeans. All of these were key innovations in the expansion of polar exploration.

The expedition's achievements were not just in technology and geography, however. The scientific results included observations on botany, zoology, meteorology and archaeology. Per Schei, the expedition geologist, was later credited with making the most significant contribution to the geological understanding of the Canadian Arctic islands of any individual before the advent of aircraft in the region.

Sverdrup had hoped to return to Norway in 1901, but the previous year he had moved *Fram* farther west, to Goose Fiord. Unfortunately, the ship had been taken so high up the fiord that she could not escape from the ice. Sverdrup was able to sail only 16 kilometres (10 miles) before having to set up for another winter. A third spring and summer were spent continuing surveys of the region, and in July 1902 the ice around *Fram* melted at last. In August, after four years in the region, Sverdrup was finally able to set sail for Norway.

LEFT A lenticular cloud dominates a beautiful, blue sky with Axel Heiberg Island in the background. In this area, Sverdrup and his party improved countless facets of skiing, sledging and polar travel.

ANTARCTIC 1901–1904
SCOTT OF THE ANTARCTIC

For several decades, Sir Clements Markham, who had supported Sherard Osborne in pushing for the government to launch the British Arctic Expedition (see pages 12–13), carried out a similar campaign to re-involve Britain in exploration of the Antarctic. Eventually Markham, by then president of the Royal Geographical Society (RGS), gathered enough support for an expedition to be organized under the auspices of the RGS and the Royal Society. Overriding the opposition of those who wanted a scientist as the expedition's leader, Markham selected instead a Royal Navy officer, Robert Falcon Scott. Markham also ensured the expedition would focus as much on geographical exploration as on science.

ABOVE *Discovery* in the ice. Scott decided that, despite a hut having being erected ashore, the ship offered a warmer, more pleasant place in which to winter.

BELOW (From left) Shackleton, Scott, and Wilson at Hut Point after their return from the gruelling Southern Journey, on which they suffered from scurvy and a number of other physical complaints.

Manned primarily by Royal Navy personnel, the British National Antarctic Expedition sailed in August 1901 on *Discovery*, the first British vessel designed specifically for scientific research. In February 1902, having traced the margins of parts of the Ross Sea, Scott landed on the Great Ice Barrier. His party inflated a balloon – and Scott became the first man to make an ascent over Antarctica; he was quickly followed by third officer Ernest Shackleton. In the following weeks, a base was established at Hut Point on Ross Island at the southern end of McMurdo Sound, where *Discovery* was frozen into a safe location. Although a hut was built ashore, it was used for work, and the men lived aboard ship.

The scientific work was initiated at once, but the primary exploratory effort only began in November 1902, when Scott and two companions – Shackleton and Dr Edward Wilson – left with a team of dogs in an attempt to reach the South Pole. They were the first to sledge deep onto the Great Ice Barrier, finding huge crevasses, steep ridges and heavy snow that made conditions far worse than imagined. Moreover, none of the three knew how to drive sledge dogs effectively, and soon they had to go into harness and begin man-hauling.

For more than a month they relayed, dragging part of the load forward and then returning for the rest, making only one mile south for every three travelled. They suffered painfully from overwork, lack of food and snowblindness, but on 30 December they managed to establish a record farthest south of 82°16' 33" S.

With supplies dwindling, the party turned for home, dragging themselves back into Hut Point on 3 February 1903 after terrible hardships including scurvy, illness and lack of food. There, Scott found that a relief ship, *Morning*, had reached McMurdo Sound. But since vast stretches of ice separated *Discovery* from open water, Scott's party was forced to winter again. When *Morning* sailed north to avoid

ABOVE An ice thermometer taken on Scott's first expedition (1901–1904). Such an item would have been part of the equipment officially being used for the scientific studies conducted under second officer Michael Barne.

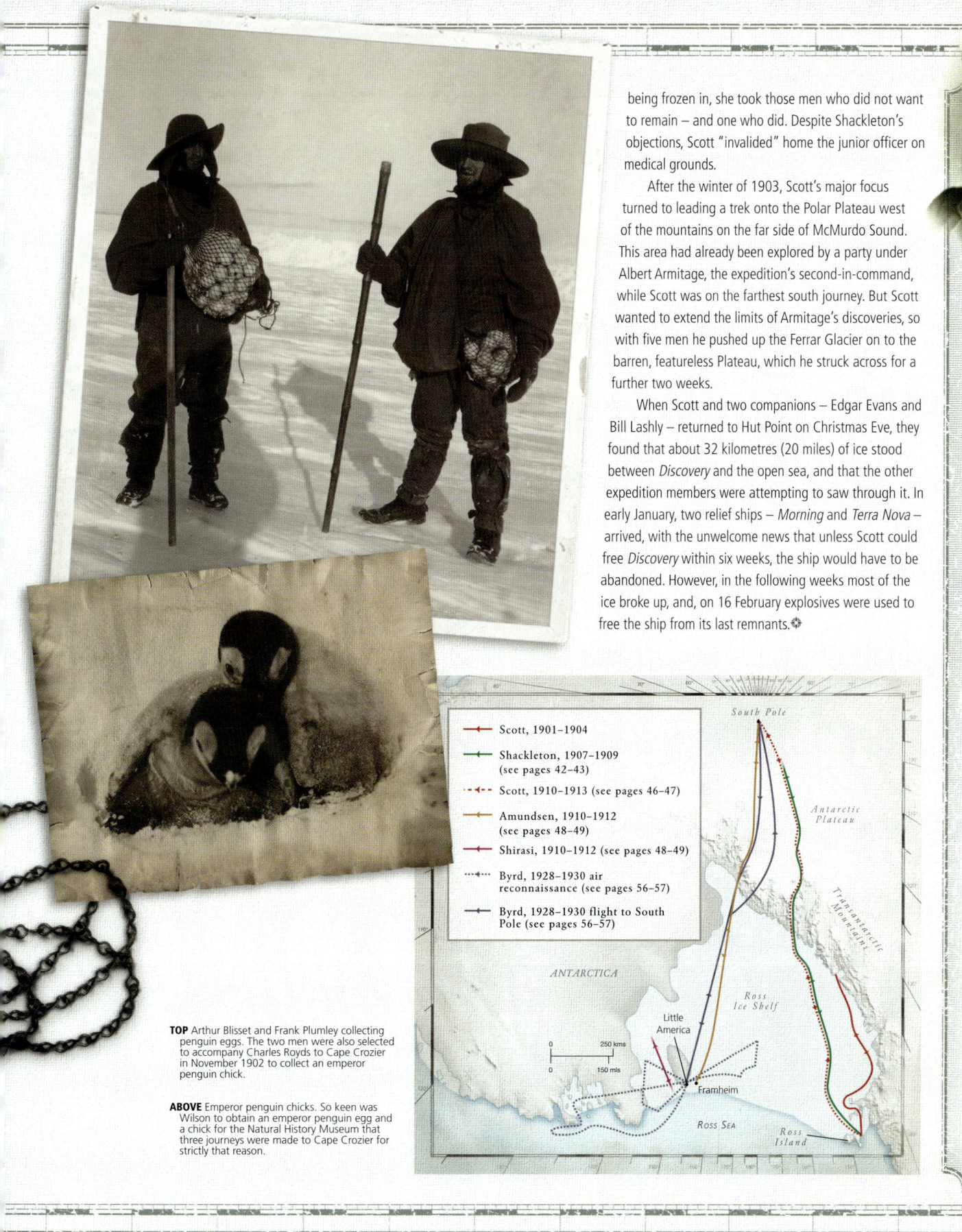

being frozen in, she took those men who did not want to remain – and one who did. Despite Shackleton's objections, Scott "invalided" home the junior officer on medical grounds.

After the winter of 1903, Scott's major focus turned to leading a trek onto the Polar Plateau west of the mountains on the far side of McMurdo Sound. This area had already been explored by a party under Albert Armitage, the expedition's second-in-command, while Scott was on the farthest south journey. But Scott wanted to extend the limits of Armitage's discoveries, so with five men he pushed up the Ferrar Glacier on to the barren, featureless Plateau, which he struck across for a further two weeks.

When Scott and two companions – Edgar Evans and Bill Lashly – returned to Hut Point on Christmas Eve, they found that about 32 kilometres (20 miles) of ice stood between *Discovery* and the open sea, and that the other expedition members were attempting to saw through it. In early January, two relief ships – *Morning* and *Terra Nova* – arrived, with the unwelcome news that unless Scott could free *Discovery* within six weeks, the ship would have to be abandoned. However, in the following weeks most of the ice broke up, and, on 16 February explosives were used to free the ship from its last remnants.

Sir Clements Markham

As a young midshipman, Clements Markham served on a Franklin search expedition under Horatio Austin, and his experiences convinced him this was the way polar expeditions should be conducted. He later worked for the India Office, gaining recognition for smuggling cinchona saplings and seeds out of Peru to help Britain establish plantations where quinine could be developed to protect against malaria. In 1863, Markham became secretary of the RGS, and in 1893 he was elevated to president. From these positions, he led his long, obsessive campaign for a British Antarctic expedition, in which he was ultimately successful.

The Attainment of the Polar Plateau

While Scott headed south over the Great Ice Barrier, Albert Armitage led a party west across McMurdo Sound and into the mountains of Victoria Land, seeking to find a route inland. After a hard ascent to a rise at about 1,520 metres (5,000 feet), the party saw below them what would be named the Ferrar Glacier. When their way was eventually blocked, they retreated down to the glacier and then followed it west. In early January, Armitage's team reached the Polar Plateau at some 2,740 metres (9,000 feet), becoming the first men ever to stand on the world's largest ice cap.

TOP Arthur Blisset and Frank Plumley collecting penguin eggs. The two men were also selected to accompany Charles Royds to Cape Crozier in November 1902 to collect an emperor penguin chick.

ABOVE Emperor penguin chicks. So keen was Wilson to obtain an emperor penguin egg and a chick for the Natural History Museum that three journeys were made to Cape Crozier for strictly that reason.

Scott, 1901–1904
Shackleton, 1907–1909 (see pages 42–43)
Scott, 1910–1913 (see pages 46–47)
Amundsen, 1910–1912 (see pages 48–49)
Shirasi, 1910–1912 (see pages 48–49)
Byrd, 1928–1930 air reconnaissance (see pages 56–57)
Byrd, 1928–1930 flight to South Pole (see pages 56–57)

RIGHT Albert Armitage, Scott's second-in-command, and leader of the first attainment of the Polar Plateau.

ANTARCTIC 1901–1904: NORDENSKJÖLD'S ADVERSITY

THE FIRST ANTARCTIC EXPEDITION TO HAVE SCIENCE PURELY AS ITS PRIMARY GOAL WAS A SWEDISH EFFORT LED BY OTTO NORDENSKJÖLD, THE NEPHEW OF THE MAN WHO HAD CONQUERED THE NORTHEAST PASSAGE (SEE PAGES 14–15). SAILING IN THE FORMER WHALER *ANTARCTIC*, COMMANDED BY CARL LARSEN, THE EXPEDITION FIRST STOPPED IN BUENOS AIRES, WHERE JOSÉ SOBRAL OF THE ARGENTINE NAVY WAS ALLOWED TO JOIN THE EXPEDITION AS A CARTOGRAPHER, WHILE THE ARGENTINE GOVERNMENT PROVIDED COAL AND FOOD IN EXCHANGE.

In early 1902, after exploring west of the Antarctic Peninsula, the expedition sailed through the strait between the top of the Peninsula and Joinville Island east of it, naming this Antarctic Sound. After being halted by the ice of the Weddell Sea, Nordenskjöld established a base at Snow Hill Island, where, with Sobral and four others, he remained to conduct scientific studies, while *Antarctic* engaged in whaling and then wintered in the Falkland Islands. Throughout the winter, Nordenskjöld's party made scientific observations, and when spring came they sledged some 320 kilometres (200 miles) south, exploring various islands and what was later named the Larsen Ice Shelf. But *Antarctic* never came, the sea froze over in February 1903, and the six men had to wait out another long winter. When spring arrived, a party sledged north to investigate, and at Vega Island they encountered three of their former companions under the command of Johan Gunnar Andersson.

Nordenskjöld was told that the previous summer *Antarctic* had completed more survey work west of the Peninsula. Larsen had then tried to take her through Antarctic Sound, but had not been able to force a passage because of heavy ice. Andersson, Samuel Duse and Toralf Grunden had then been landed at Hope Bay, near the tip of the Peninsula, so that they could reach Nordenskjöld by sledge, while Larsen

ABOVE An ice barrier near Snow Hill Island. In recent years, with the breaking up of parts of the Larsen Ice Shelf, the physical nature of the western Weddell Sea region has radically changed.

FAR RIGHT *Antarctic* in the process of sinking, having been crushed in the ice. It took the ship's party 16 days to reach land at Paulet Island.

BELOW Scientific work near the Larsen Ice Shelf. One of the expedition's most important scientific accomplishments was finding a number of fossils, including the bones of an ancient, giant penguin.

34

Map legend:
- Nordenskjöld's sledge journey, 1902
- Andersson's journey from Hope Bay, 1903
- Andersson & Nordenskjöld's return together, 1903
- Larsen's journey from Paulet Island, 1903
- *Antarctic* sinks

BELOW The relief ship *Uruguay*. The Argentine government viewed the effort as a race with France and Sweden and as a statement of its national significance.

left these quarters in September and successfully reached Nordenskjöld. But where was *Antarctic*?

The previous year, after dropping off Andersson, Larsen had taken the ship around Joinville Island, but while trying to go south, she was nipped in the ice. The pressure caused a major leak. An attempt to beach her on Paulet Island was prevented by heavy ice, and when the pumps could no longer keep up with the leak, the crew put everything they could on the ice, and watched as *Antarctic* sank 40 kilometres (25 miles) east of Paulet. For more than two weeks, they slowly made their way towards Paulet Island, rowing through the shifting ice, ferrying stores from floe to floe by whaleboat and camping uneasily on the floes. At the end of February 1903, they finally struggled ashore, where they promptly built a hut and killed about 1,100 penguins to help them get through the winter. During that winter, the Norwegian sailor Ole Wennersgaard died of heart disease, and his colleagues could only bury him in a snowdrift because of the hardness of the ground.

In October, Larsen and a party of five set out for Hope Bay, only to discover after a difficult journey that Andersson had left for Snow Hill Island. Larsen and his men followed, but finding that the ice in Erebus and Terror Gulf had broken up, they were forced to row for 200 kilometres (125 miles). When they reached Snow Hill, they found not only Nordenskjöld's and Andersson's parties, but the Argentine relief ship *Uruguay*. All the men were soon loaded aboard the ship, which sailed to Paulet to collect the others. After a stop at Hope Bay to pick up Andersson's fossil collection, they steamed north to Buenos Aires.

tried to sail north around Joinville Island and back south in what he hoped would be better ice conditions.

Andersson's party had not been able to cross an area of open water, so had returned to Hope Bay, assuming Larsen would be able to reach Snow Hill Island and would then return for them, as agreed. But *Antarctic* did not come back, so they built a hut, where they spent a miserable winter. They

RIGHT The base at Snow Hill Island. Perhaps South Georgia did not seem so remote to Larsen after leaving Nordenskjöld here.

THE FOUNDING OF WHALING ON SOUTH GEORGIA

In 1902, after leaving Nordenskjöld's party at Snow Hill Island and before going to the Falklands, Carl Larsen took *Antarctic* to South Georgia, where he had seen rorquals the previous decade. There he visited a small bay named Grytviken and realized that it would make an ideal location for a whaling station. When the members of the expedition were later taken to Buenos Aires, Larsen persuaded local financiers to back his idea. Whaling operations started at Grytviken in December 1904, and South Georgia soon became the centre of the Southern Ocean's whaling industry.

THE RELIEF OF NORDENSKJÖLD

By 1903, anxiety had grown in official circles about the fate of Nordenskjöld's expedition. In Sweden, a relief operation was organized under Captain Olof Gyldén. At the same time, an Argentine effort was launched under Lieutenant Julián Irizar. Both governments wanted to reach Nordenskjöld first, and the French became involved, too, when Jean-Baptiste Charcot offered to divert his expedition to search for the missing men. In the end, Irizar arrived before Gyldén – who reached Snow Hill Island to find everyone gone – and in time for Charcot to redirect his expedition to scientific and geographical aims.

ANTARCTIC 1901–1904
THE SCIENTISTS HEAD SOUTH

LEFT Erich von Drygalski, leader of the German expedition. He never returned to the Antarctic, but did take part in an expedition to Svalbard, before devoting much of his later research to anthropogeography.

THE NOTED SCIENTIST GEORG VON NEUMAYER HAD BEEN PROMOTING THE IDEA OF A GERMAN ANTARCTIC EXPEDITION FOR MORE THAN THREE DECADES WHEN, IN THE CLOSING YEARS OF THE NINETEENTH CENTURY, THE PROMISE OF GOVERNMENT FUNDING FINALLY MADE HIS VISION A REALITY. NEUMAYER AND HIS COLLEAGUES ON THE OFFICIAL ANTARCTIC COMMITTEE CHOSE ERICH VON DRYGALSKI, PROFESSOR OF GEOGRAPHY AND GEOPHYSICS AT BERLIN UNIVERSITY AND LEADER OF TWO EXPEDITIONS TO GREENLAND, AS THE LEADER OF WHAT WAS OFFICIALLY NAMED THE GERMAN SOUTH POLAR EXPEDITION.

By 1901, the specially built research vessel *Gauss* was ready, and the expedition headed towards a little-known area of Antarctic coast, with five scientists, five naval officers, and a crew of 22. Drygalski's plans included the taking of observations simultaneously with those made by Scott's and Nordenskjöld's parties (see pages 32–35).

After stopping at Iles Kerguelen, *Gauss* made her way towards the Antarctic coast near 90°E, but in February 1902, she was beset in the ice some 85 kilometres (53 miles) off shore. For the next year, while *Gauss* drifted off what they named Kaiser Wilhelm II Land, the party conducted comprehensive scientific studies, made ascents in a tethered hydrogen balloon and carried out sledge trips toward the continent. There they discovered – in the midst of a generally ice-covered coast – an ice-free, extinct volcano, which they named Gaussberg.

Finally, in March 1903, the ship escaped from the ice, but rather than immediately turn back north, Drygalski made attempts to go farther south – to the west of where *Gauss* had been imprisoned – in order to trace the continent's unknown contours. These

BELOW *Gauss* trapped in the ice in early summer, 1902, after a heavy snowstorm. It was another three months before the ice began to lose its ferocious grip, and a month after that until the ship was totally free.

ABOVE Members of the German expedition ready to begin the journey towards the extinct volcano Gaussberg. With no base ashore, the ship seemed even more than usual a haven of safety and comfort.

RIGHT *Scotia* temporarily halted in the ice. The men on skis and the lookout in the crow's nest suggest that efforts were being made to find a lead in which to advance, as does the fact that a number of sails are still set.

final attempts were unsuccessful, however, and the expedition was forced to turn north, having gained enough scientific data to fill 20 large volumes and two atlases.

At the same time that Drygalski was trapped in the ice, another expedition was heading south, led by the world's foremost polar scientist. William Speirs Bruce had first come to the Antarctic as surgeon on *Balaena* during the Dundee whaling expedition of 1892–1893, during which he had realized the enormous value of scientific research in the polar regions. Several years later, he served as a naturalist on the Jackson-Harmsworth Expedition to Franz Josef Land (see page 23), and he then returned to the Russian Arctic as a naturalist with the industrialist Andrew Coats in 1898. After an expedition to Spitsbergen with the great oceanographer Prince Albert of Monaco, Bruce applied to join the British National Antarctic Expedition (see pages 32-33), but was rejected by Clements Markham.

Bruce then determined to lead his own expedition, one that would be purely Scottish and that, while concentrating on oceanographic study of the Southern Ocean, would also include the compilation of meteorological, biological, topographical and physics data. With a large donation from the philanthropic brothers Andrew and James Coats, the former whaler *Hekla* was converted to an oceanographic research vessel, which Bruce named *Scotia*. The Scottish National Antarctic Expedition sailed with an extensive scientific staff and a crew made up of seasoned whalers.

In February 1903, they landed in the South Orkney Islands, and then proceeded into the Weddell Sea as far south as 70°21′ S. However, not wishing to be trapped by the ice in an area without great scientific opportunities, Bruce retreated north, back to the South Orkneys. In March the party anchored off of Laurie Island, where *Scotia* was frozen in until November. There they built Omond House, which served both as a meteorological observatory and their living quarters.

In early 1904, Bruce and his party made another voyage into the Weddell Sea, discovering along its western boundary a hitherto unknown area they named Coats Land. However, despite following the coastline and ultimately reaching a latitude of 74°1′ S, they were unable to land. Then a sudden blizzard caused them to be caught by a tricky section of ice and, facing the possibility of becoming trapped, Bruce decided to reverse course as soon as possible. In March, he turned north, heading for home having completed the most successful oceanographic voyage of any Antarctic expedition.

LEFT Gilbert Kerr, the official piper of the Scottish expedition, performing in full Highland dress. The emperor penguin had to be tied to a cooking pot packed with snow to prevent it from wandering off.

GEORG VON NEUMAYER

Like Clements Markham in Britain, Georg von Neumayer was ultimately successful after years of urging his government to participate in Antarctic research. A talented scientist with a background in meteorology, magnetism, and hydrology, Neumayer founded a geophysical observatory in Melbourne in 1857, intending it as a base for later Antarctic work. After returning to Germany, he served as hydrographer to the navy, head of the German Naval Observatory, and president of the German Meteorological Society. He was a key figure in planning the International Polar Year (1882–1883), and was the guiding force behind the preparations for Drygalski's Antarctic expedition.

OMOND HOUSE

In late 1903, Bruce, hoping to extend the scientific data his expedition had already compiled, invited the Argentine government to take over and maintain Omond House, the meteorological station founded earlier that year on Laurie Island, and the tiny adjacent magnetic observatory named Copeland House. The Argentines readily agreed, and the next February the transfer took place, although Bruce's colleague Robert Mossman remained in charge of the base for another year. The Argentine station was renamed Orcadas in 1951, and is now the oldest continually manned base in Antarctica.

BELOW Omond House, as it appeared during Bruce's expedition. Named for Scottish meteorologist Robert Omond, who advocated scientific studies in Antarctica, it was built of more than 100 tons of stone.

ANTARCTIC 1903–1910

CHARCOT'S TWO VOYAGES

THROUGHOUT HIS YOUTH, JEAN-BAPTISTE CHARCOT DREAMED OF A LIFE AT SEA, AND HOPED TO JOIN THE FRENCH NAVY. HOWEVER, HIS FATHER, A RENOWNED NEUROLOGIST, INSISTED HIS SON STUDY TO BECOME A DOCTOR, AND CHARCOT FOLLOWED HIS WISHES. BUT THE DEATHS OF HIS PARENTS LEFT CHARCOT BOTH FREE AND WITH A FORTUNE, AND HE SOON ABANDONED THE MEDICAL PROFESSION. IN THE FOLLOWING YEARS, HE PURCHASED A SERIES OF PROGRESSIVELY LARGER YACHTS – EACH OF WHICH HE NAMED *POURQUOI PAS?* – AND MADE CRUISES FARTHER AND FARTHER AFIELD.

After a trip to the Arctic, Charcot commissioned the building of a ship, named *Français*, that he hoped to take to northern Greenland. The costs of the ship proved too much for his budget, however, and Charcot eventually economized by having an underpowered engine installed, which would later prove a major problem in the ice.

In 1903, when reports indicated that Otto Nordenskjöld's party had disappeared in the Antarctic (see pages 34–35), Charcot decided to go south to search for them. However, when he arrived at Buenos Aires, Charcot found out that Nordenskjöld's party had been relieved by the Argentines. So he turned to the west of the Peninsula instead, to explore and chart its unknown reaches and to conduct a comprehensive scientific programme.

In early March 1904, after *Français* had suffered serious engine problems, Charcot decided to winter at Booth Island. His thorough planning allowed his party to have one of the most comfortable and productive winters of any Antarctic expedition. In the spring, after carrying out months of scientific studies, the expedition continued the survey of the Graham Land coast and the islands off it, including passing inland of Adelaide Island. But on 15 January 1905, *Français* struck a submerged rock and was badly holed. Temporary repairs were made at Port Lockroy off Wiencke Island, but the ship was so damaged that Charcot was forced to return to Buenos Aires, where the Argentine government gladly purchased her. Despite the expedition being thus curtailed, more than 960 kilometres (600 miles) of coastline had been charted, and detailed scientific results obtained.

When Charcot returned to France, he discovered that his wife had divorced him on the grounds of desertion. But throughout France he had become a hero, and government and newspaper sponsorship allowed him to build a larger, more powerful ship, which he again named *Pourquoi-Pas?*, and to plan another expedition to extend his work in the Antarctic Peninsula region.

The expedition left Le Havre in August 1908, and by December he had reached his old wintering site at Booth Island. Shortly thereafter, he discovered a sheltered

ABOVE Final preparations are made for the departure of *Français* from Le Havre in August 1903. The ship had to return after a sailor was killed when struck by a hawser. The expedition left again 12 days later.

LEFT *Pourquoi-Pas?* arrives at Rouen in June 1910. Not only did Charcot survey thousands of kilometres of Antarctic coastline, but he did so with such remarkable accuracy that some of his charts were not improved upon for decades.

ENCLOSURE

Part of the plan of *Pourquoi-Pas?*. The experience of having *Français* badly damaged led Charcot to have his new ship built stronger to withstand the ice, but it, too, received significant damage in the Antarctic.

harbour on nearby Petermann Island, which he named Port Circumcision. Charcot's bad luck continued, however, and in January 1909 *Pourquoi-Pas?* ran aground, although it was refloated hours later. Once again, significant damage had occurred, but Charcot decided it was not serious enough to halt his exploration.

Soon thereafter, the party carried out a meticulous mapping of Adelaide Island, including discovering a large bay nearby – named Marguerite Bay after Charcot's new wife. The expedition then approached Alexander Island before lack of coal and increasing ice forced a retreat to Port Circumcision. There, Charcot had steel hawsers placed across the harbour's narrow entrance to prevent icebergs from entering and further damaging the ship.

After a winter equally as successful as that of the previous expedition, journeys were made from the base to the surrounding areas. When *Pourquoi-Pas?* was freed in November 1909, Charcot took the ship to Deception Island, where he was able both to obtain coal and to have it surveyed. He then headed back to the Peninsula, where, amongst other achievements, the expedition discovered Charcot Island, named for the explorer's father. In mid-January, the party sighted Peter I Øy, which had not been seen since Fabian von Bellingshausen's expedition nine decades before. After running west along the edge of the continent for another week, Charcot finally headed back to South America, having charted some 2,000 kilometres (1,250 miles) of coastline and secured enough scientific data to fill 28 volumes of reports.

ABOVE Members of Charcot's first Antarctic expedition hold aloft a pair of petrels on *Français*. When two naturalists quit the expedition in Buenos Aires, Jean Turquet joined as zoologist.

ABOVE After returning from the south, Dallmann made a series of voyages (1877–1883) in hopes of establishing a regular trade route from Germany to Siberia.

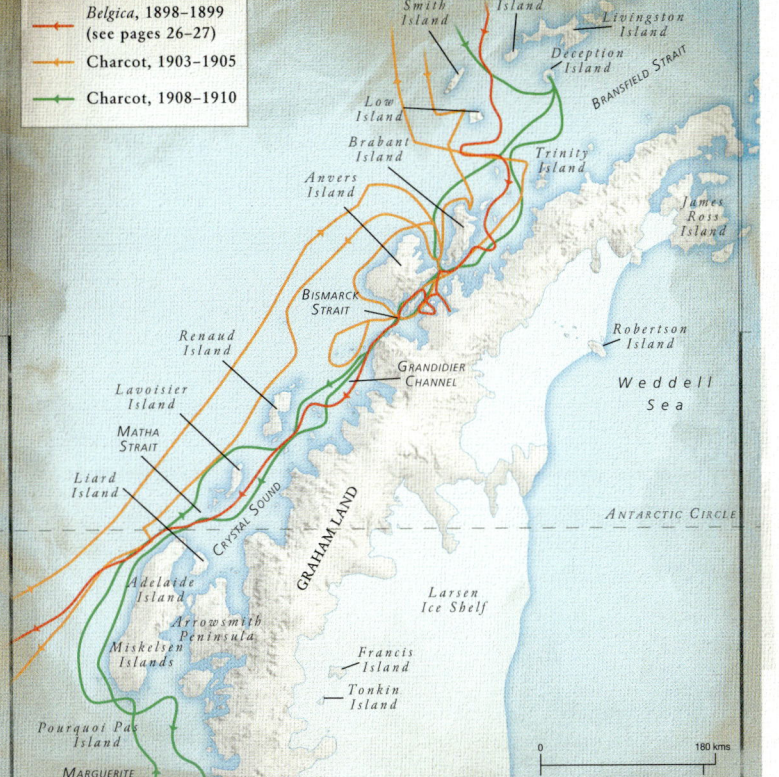

BELOW Charcot (centre, with white beard) showing guests around *Pourquoi-Pas?* in 1936, the year of his death. On the right is the next great French polar figure, Paul-Émile Victor.

EDUARD DALLMANN'S ANTARCTIC VOYAGE

In 1873, a Bremerhaven whaling company sent Eduard Dallmann to the Antarctic Peninsula to investigate whether "black whales" reported by James Clark Ross could be southern right whales. Between November 1873 and February 1874, Dallmann explored the little-known coastal waters west of the Peninsula, discovering, naming and charting many features, including Bismarck Strait, Booth Island, Petermann Island and others that were later renamed (such as Neumayer Channel). Although no right whales were sighted, numerous rorquals were, and the expedition proved financially successful as Dallmann brought back fur seal pelts and seal oil.

CHARCOT'S LATER CAREER

Although Charcot never led another Antarctic expedition, he continued his adventures in the ice. After the First World War, he began a series of yearly voyages in *Pourquoi-Pas?* to the Arctic, North Atlantic or other areas for research. In 1925, he sailed to Greenland, after which he became a regular visitor, including helping establish a French research station at Scoresby Sund. In 1936, he was returning to France from Greenland and Iceland when *Pourquoi-Pas?* was caught in a violent storm and foundered. Charcot and all but one of his crew went down with the ship.

Belgica, 1898–1899 (see pages 26–27)
Charcot, 1903–1905
Charcot, 1908–1910

ARCTIC 1903–1906
THE NORTHWEST PASSAGE

BELOW *Gjøa* during the summer at Gjøahavn. After being the first ship to complete the Northwest Passage, she was left to decay for years, but now sits proudly, adjacent to the Frammusset in Oslo.

BELOW Amundsen with two sledge dogs. The lessons he learned from the Netsilik Inuit about living and working in the polar regions proved invaluable in his South Pole expedition.

WHEN ROALD AMUNDSEN WAS 15, HE READ JOHN FRANKLIN'S ACCOUNTS OF HIS TWO LAND EXPEDITIONS IN THE FAR NORTH OF CANADA (1819–1822 AND 1825–1827), AND THEREAFTER THE YOUNG NORWEGIAN'S SOLE AMBITION WAS TO BECOME AN ARCTIC EXPLORER. HIS FIRST GREAT GOAL WAS TO NAVIGATE THE NORTHWEST PASSAGE, THE COMPLETION OF WHICH HAD ELUDED EXPLORERS FOR HUNDREDS OF YEARS. AMUNDSEN SPENT MOST OF HIS TWENTIES PREPARING HIMSELF FOR THIS CHALLENGE, INCLUDING EARNING HIS MASTER'S CERTIFICATE FOR SAILING VESSELS, BECOMING AN EXPERT SKIER AND SERVING AS MATE ABOARD *BELGICA* DURING THE FIRST WINTERING IN THE ANTARCTIC (SEE PAGES 26–27).

By 1903, he had obtained the sponsorship of King Oscar II, consulted former veterans of the search for the Northwest Passage, and conducted extensive studies into magnetic observations – so as to be able to determine the exact location of the North Magnetic Pole. Amundsen finally felt prepared for his attempt on the Passage, and he selected a party of six men to accompany him, setting out in June aboard the 47-ton former fishing ship *Gjøa*.

Upon reaching northwest Greenland, Amundsen purchased 10 Inuit dogs, some sledges and a few kayaks. He then took *Gjøa* west through Lancaster Sound and made a landing at Beechey Island, Franklin's last wintering site. Observations indicated that the North Magnetic Pole lay to the south, so Amundsen sailed down little-known Peel Sound and along the west coast of Somerset Island and Boothia Peninsula before temporarily running aground near Matty Island. Having refloated the ship, he found a protected bay in which to winter on the southeast coast of King William Island. Within days, *Gjøa* was frozen in at what Amundsen named Gjøahavn.

The Norwegians were soon joined by parties of Netsilik Inuit, who in the following months taught them how to build igloos, live and work in fur clothing and sledge in the coldest of temperatures. Meanwhile, Amundsen made detailed studies of the Netsilik and their customs. In the spring of 1904, Amundsen headed towards the point where James Clark Ross had first reached the North Magnetic Pole. The party's measurements indicated that the Pole was no longer located there, proving that over time it migrated over the surface of the Arctic. On another sledging trip in August, Godfred Hansen and Helmer Hanssen discovered skeletal remains of two men from Franklin's expedition.

An early onset of winter conditions in September 1904 saw the ship again frozen in, and the party battened down for a second year at Gjøahavn. The next spring, Hansen and Peder Ristvedt spent two months charting the last unexplored parts of the Victoria Island coast.

RIGHT The late winter in Gjøahavn, showing how the tiny expedition ship almost disappeared under the accumulated snow and ice of the high Arctic.

BELOW Amundsen and the crew of *Gjøa* at Nome, Alaska. Amundsen continued to San Francisco via steamship, and Wiik had died in the Arctic, so two new men were added to help crew the little ship.

In August 1905, *Gjøa* finally sailed west, passing through unknown waters, some of them so shallow that no previous ship in the region would have been able to navigate them. Off Nelson Head, on Banks Island, they met the San Francisco whaler *Charles Hanson*, the captain of which – having heard of Amundsen from the Norwegian consul – congratulated him on his magnificent achievement.

However, the expedition was not finished yet. Halted by heavy ice near Herschel Island, west of the Mackenzie River delta, *Gjøa* was forced to winter a third time, near an American whaling fleet. During the winter, Amundsen joined two Inuit and one of the whaling captains in sledging approximately 800 kilometres (500 miles) to Eagle City, Alaska, to break the news of his success via telegraph. He sent his story to Fridtjof Nansen, but the message was read en route and its contents leaked to the press. When Amundsen returned to *Gjøa* in March, he found Gustav Wiik, the expedition's most dedicated scientist, seriously ill; Wiik died a few days later.

In July 1906, *Gjøa* was finally able to advance again, and the next month she rounded Point Barrow and passed through Bering Strait, reaching San Francisco in October. Anxious to get back to Europe, Amundsen and his party left his heroic little ship behind, and returned to Norway by a faster mailboat. *Gjøa* remained, generally forgotten, in San Francisco, and it was not until 1972 that she was finally brought back to Oslo and given a place of honour.

BELOW In the saloon of *Gjøa*, March 1904. Seated in Netsilik reindeer fur anoraks are (left to right): Amundsen, Helmer Hanssen, who later accompanied him to the South Pole and Peter Riestvedt.

LEFT Anthony Fiala first went to the Arctic as photographer on an earlier expedition funded by William Ziegler, which had proved a total fiasco.

THE FIALA-ZIEGLER EXPEDITION

While Amundsen was in the Canadian archipelago, a highly publicized effort in Arctic Russia, financed by American millionaire William Ziegler, proved considerably less successful. Led by Anthony Fiala, the expedition had problems from the start, when its ship, *America*, was crushed in the ice off Franz Josef Land in November 1903. In the next year and a half, Fiala three times headed north for the Pole, only to return each time within a few days, never bettering 82° N. In the summer of 1905, a relief ship finally rescued the members of the expedition, who had achieved virtually nothing.

AMUNDSEN AND THE NORTHEAST PASSAGE

Although Amundsen is best remembered for being the first man to attain the South Pole and to navigate the Northwest Passage, one of his other great achievements was passing through the Northeast Passage. Before going south, Amundsen planned an Arctic drift like that of *Fram*, and in 1918 that expedition finally sailed in *Maud*, named after the Norwegian queen. It was perhaps Amundsen's least successful effort, and in 1920, after two winters trapped in the ice north of Siberia, efforts to force *Maud* into the drifting pack failed. The expedition therefore continued west to Alaska, becoming only the third venture to navigate the Northeast Passage.

RIGHT Expedition members on *Maud* wave as they leave Seattle, from where they again entered the ice to attempt a polar drift.

SHACKLETON'S FARTHEST SOUTH

ANTARCTIC 1907–1909

IN THE YEARS AFTER BEING "INVALIDED" HOME FROM THE *DISCOVERY* EXPEDITION BECAUSE HE WAS SUFFERING FROM SCURVY ON THE RETURN FROM THE FARTHEST SOUTH (SEE PAGE 33), ERNEST SHACKLETON DETERMINED TO LEAD HIS OWN EFFORT TO REACH THE SOUTH POLE. IN 1907, HE RECEIVED A LOAN FROM SCOTTISH INDUSTRIALIST WILLIAM BEARDMORE, ALLOWING HIM TO PURCHASE THE TINY SEALER *NIMROD*, PUT TOGETHER A SHORE PARTY AND HASTILY DEPART ON WHAT HE GRANDLY NAMED THE BRITISH ANTARCTIC EXPEDITION.

ABOVE *Nimrod* in the ice. The former sealing vessel was named after the grandson of Ham and great-grandson of Noah in the Book of Genesis. The original Nimrod "was a mighty hunter before the Lord".

To honour a promise to Robert Falcon Scott to leave "his" sector near McMurdo Sound alone, Shackleton initially attempted to establish a base on the Great Ice Barrier. When this proved impossible, Shackleton set up at Cape Royds, on Ross Island, about 29 kilometres (18 miles) north of Hut Point. Unfortunately, the sea ice covering McMurdo Sound broke up shortly thereafter, eliminating access to the Barrier for the party of 15. In order to keep his men occupied during the period before the sea refroze, Shackleton ordered an ascent of Mount Erebus, the active volcano on Ross Island. After an extended struggle, five of the six-man party reached the summit, the first ever to do so.

The next spring, Shackleton sent out three exploring parties. The Western Party conducted geological studies in the mountains west of McMurdo. Concurrently, the Northern Party – consisting of T W Edgeworth David, Douglas Mawson and Alistair Mackay – made a gruelling, four-month man-hauling journey of some 2,000 kilometres (1,250 miles). After following the coast of Victoria Land to the north, they took a tortuous path up a series of glaciers to the Polar Plateau, which they slogged across towards the South Magnetic Pole. In January 1909, they became the first men ever to reach the region of that Pole, but with their time to meet *Nimrod* nearing, they had to race back to the coast. They were only saved after John King Davis, the ship's first mate, insisted that the search for them, which the captain had terminated, visit one last site – and mercifully, they were there.

Meanwhile, Shackleton and three companions – Frank Wild, Jameson Adams and Eric Marshall – set off towards the expedition's main goal: the South Pole. A motor-car that Beardmore had donated proved essentially useless, so four Manchurian ponies were used to haul equipment and supplies. Within a month of leaving Cape Royds, the party surpassed Scott's farthest south (82° 16'33" S) (see pages 32), but despite this success, the ponies soon proved unsuitable for the glacial conditions, and three of them were eventually put down. Having crossed the Barrier, the men found the vast Transantarctic Mountains blocking their path, so they set off through them up the unknown Beardmore Glacier. Four days later, Socks, the final pony, disappeared down a deep crevasse, and the men had to do all of the hauling themselves. Nevertheless, despite short rations,

ABOVE A small knife that belonged to Bertram Armytage, the leader of Shackleton's Western Party and one of three Australians on the expedition. Sadly, Armytage committed suicide in Melbourne in 1910.

LEFT Atop Mount Erebus. Jameson Adams, T W Edgeworth David, Alistair Mackay, Eric Marshall and Douglas Mawson reached the summit after leaving Philip Brocklehurst behind in a sleeping bag with frostbitten toes.

ABOVE At the start of the Southern Journey, the four ponies – Socks, Grisi, Quan and Chinaman – each pulled a sledge.

PONIES TOWARDS THE POLE

None of the men on Scott's farthest south – Scott, Shackleton or Edward Wilson – was a proficient dog-driver, so Shackleton was sceptical about the benefits of dogs for Antarctic travel. When planning his expedition, he followed the recommendation of Frederick Jackson to use ponies, as Jackson had in Franz Josef Land. Shackleton brought 10 Manchurian ponies south, but six died aboard *Nimrod* or after reaching Cape Royds. Although the remaining four helped the Southern Party cross the Barrier, they proved unsuitable for the conditions, suffering terribly in the cold, requiring bulky fodder to be carried and frequently breaking through snow coverings with their sharp hooves.

THE PROF: T W EDGEWORTH DAVID

T W Edgeworth David, known affectionately as "The Prof", was one of the world's most respected geologists when he asked Shackleton if he could temporarily come south with him to see the Antarctic. When Shackleton agreed, David helped him obtain a large grant from the government of Australia, which in turn allowed Shackleton to hire Douglas Mawson. David later decided to stay for the entire time in the Antarctic, and he and Mawson formed an impressive scientific component. Upon returning to Australia, David became a national hero, and was later knighted for contributions to science.

RIGHT Despite turning 51 during the expedition, David still participated in the unrelentingly harsh man-hauling on the Northern Party.

they slowly crept up the 200-kilometre-long (125-mile) glacier, finally reaching the Polar Plateau in the very heart of Antarctica at the end of December 1908.

But with winds of up to 145 kilometres (90 miles) per hour blowing in their faces; the temperatures dropping to –29° C (–20° F); suffering from altitude sickness and dehydration; and having basically run out of food, they were finally forced to end their march south at 88°23′ S, just 180 kilometres (97 geographical miles) from the Pole.

ABOVE The farthest south, at 88°23′S on 9 January 1909. Marshall took the picture of (from left): Adams, Wild and Shackleton.

BELOW The Southern Party safely back aboard *Nimrod*. From left: Wild, Shackleton, Marshall and Adams.

The journey home was even more precarious, and more than once they reached the depots they had laid on the way out in the nick of time. But somehow they struggled on, willing themselves down the Beardmore and across the Barrier so they could meet *Nimrod* before, following Shackleton's orders, she sailed north on 1 March. When Marshall collapsed 53 kilometres (33 miles) from Hut Point late on 27 February, Shackleton left him with Adams and continued on with Wild, marching 30 hours straight to catch the ship. Then, with help from Mawson and others, Shackleton returned to the Barrier for Marshall and Adams before finally heading towards New Zealand, where he would find himself an international hero.

BELOW One of the medals awarded to the members of the British Antarctic Expedition. The entire shore party also received the Polar Medal in gold, while a number of the officers and crew of *Nimrod* were awarded it in silver.

THE GREAT CONTROVERSY

ARCTIC 1907–1909

DURING THE FIRST DECADE OF THE TWENTIETH CENTURY, ROBERT E PEARY WAS FIRMLY ESTABLISHED AS THE UNITED STATES' GREATEST POLAR EXPLORER, HAVING MADE FIVE EXPEDITIONS TO GREENLAND BEFORE HE TURNED HIS EYES TOWARDS THE NORTH POLE. FOR YEARS HE HAD RECEIVED SPONSORSHIP AND BACKING FROM THE NATIONAL GEOGRAPHIC SOCIETY, THE EXPLORERS' CLUB AND THE PEARY ARCTIC CLUB – FORMED FROM AN EXCLUSIVE BAND OF NATIONALISTIC MILLIONAIRES, WHO BACKED HIS GOAL TO RAISE THE AMERICAN FLAG AT THE POLE BEFORE ANYONE ELSE ATTAINED IT.

In April 1902, towards the end of the four-year expedition on which he had encountered Otto Sverdrup, Peary reached 84°17′ N, an American record, but well short of the latitudes attained by Fridtjof Nansen and the Duke of Abruzzi (see pages 22 and 30–31). But Peary was soon back in the north, and in April 1906 he claimed a farthest north of 87°06′ N. This was still not good enough, however, as Peary, his supporters and the American public were disappointed with anything less than the Pole itself. So Peary determined to make one last attempt.

Meanwhile, the North Pole was also calling to Dr Frederick A Cook, who had first travelled to the Arctic on one of Peary's expeditions in 1891 (see pages 26–27). He later gained an international reputation for his role on *Belgica* in the first Antarctic wintering, and in 1906 claimed the first ascent of Mount McKinley, the highest peak in North America.

In 1908, Peary launched his final expedition, and, after wintering on Ellesmere Island, in February 1909 he headed north across the sea ice of the Arctic Ocean. On 1 April, Peary's last support party, headed by his ship's captain, the renowned Bob Bartlett, turned back at 87°47′ N, leaving Peary, his manservant Matthew Henson and four Eskimos to continue. According to Peary, they reached the Pole on 6 April and then made record speed on their return to the ship. Despite this claim of success, during the party's long journey from the Arctic back to the US, Peary found that his worst nightmare had come true.

In the first days of September 1909, upon arriving in Copenhagen from Greenland, Cook had stated that in April 1908 he had reached the North Pole after trekking north from Axel Heiberg Island with two Eskimos. However, he claimed, he had then missed his depot on his return and been forced to winter on Devon Island, thereby delaying his voyage to Greenland. Cook's story had been immediately purchased by James Gordon Bennett of *The New York Herald*, which gave it enormous coverage. Although there were some who doubted Cook's claims, in general they were accepted by most experts – it had always been the case that the geographical community had not demanded proofs of the honesty of such statements.

However, upon hearing of Cook's claims, Peary immediately stated that he had reached the Pole and that Cook was a fraud. Peary's cause was promptly taken up by *The New York Times*, which had sponsored his expedition, as well as by the National Geographic Society and his other powerful supporters. In the following months an international debate – fuelled by the major newspapers

ABOVE Peary in a posed shot with "Eskimo dogs" aboard his ship *Roosevelt*. In reality, the ship seemed overrun with dogs, since more than 200 were brought aboard during his attempts on the North Pole.

ABOVE Matthew Henson was a clerk in a furrier's store in 1887 when Peary hired him as his valet. He spent the next 22 years with Peary, playing a significant role in all of Peary's expeditions.

Peary's Claim Disputed

Even today, polar historians question Peary's claims to have reached the North Pole, with their doubts arising for several reasons. First, after Bartlett was suddenly sent back, no one was left in the polar party who could affirm Peary's reckonings of having reached the Pole, as neither Henson nor the four Eskimos could make or verify the necessary measurements. Second, Peary's speeds once he was away from Bartlett increased at such an astonishing rate as to be highly questionable, particularly as it would have meant travelling faster than anyone ever previously had over sea ice. And third, not only did Peary not record in his diary that he had reached the Pole – leaving that page blank – but he did not indicate to Henson that they had. It was not until after he learned of Cook's claim that he finally insisted that he had actually reached the Pole

Cook's Mount McKinley Claim

In 1906, Cook was a member of a party exploring central Alaska. After the expedition broke up, Cook returned inland with one assistant, Ed Barrill. When they returned several weeks later, Cook claimed to have conquered Mount McKinley. Belmore Browne, the expedition's artist, was sceptical and questioned Barrill, who refused to confirm Cook's claim. Browne and his colleague Professor Herschel Parker publicly stated that Cook had faked his climb, but were forced to stop such statements when Cook threatened a libel case. But in 1909, Barrill signed an affidavit swearing they had never climbed the mountain, and the next year Browne and Parker returned from an expedition with photographs they claimed showed the peak Cook had actually climbed – it was not Mount McKinley. These and other negative claims ruined Cook's reputation.

ABOVE Peary's group struggles over a large pressure ridge in the Arctic Ocean. Such ridges are formed when large, separate masses of sea ice are forced together, with nowhere for the ice at the edges to go but up.

LEFT Peary's photograph of the members of his party cheering the Stars and Stripes while supposedly at the North Pole on 7 April 1909. From left: Ooqueah, Ootah, Matthew Henson, Egingwah and Seegloo.

RIGHT Peary in full polar garb. He had such pictures very carefully prepared to establish his public image.

supporting the different explorers – raged about which of the men, if either of them, had reached the North Pole.

Although the public initially favoured Cook, it ultimately became accepted that he was lying when it was shown that his prior claim to have made the first ascent of Mount McKinley was a fabrication. The public, the scientific societies and even the US Congress then made the curious distortion of logic that if Cook hadn't reached the top of Mount McKinley, then he hadn't reached the Pole, and if he hadn't reached the Pole, then Peary had. Thus Cook's claims were discredited, and, by the end of 1910, Peary was officially acknowledged by most geographical societies and by Congress as the first man at the North Pole.

Yet it is a debate that has not disappeared even a century later. Both men still have avid supporters, while in recent years most serious polar historians have come to the conclusion that, in fact, neither man reached the Pole.

ANTARCTIC 1910–1913
SCOTT'S LAST EXPEDITION

Shortly after Ernest Shackleton's return from the Antarctic, Robert Falcon Scott announced his plans for an expedition that, he stated, would both attain the South Pole and conduct a broad scientific programme. The expedition ship *Terra Nova* – the same involved in the relief of Scott's first expedition – left Cardiff on 15 June 1910, and about four months later arrived at Melbourne. There, awaiting Scott, was a message from the Norwegian explorer Roald Amundsen. "Beg leave to inform you", it stated, "Fram proceeding Antarctic". It was the first indication that Scott's attempt on the Pole had become a race.

BELOW One of H G Ponting's most beautiful pictures was taken from inside an ice grotto, which the photographer described as "a lovely symphony of blue and green". Griffith Taylor and Charles Wright are shown in the grotto's entrance, while *Terra Nova* is framed in the distance.

In January 1911, unable to reach Hut Point on Ross Island because of ice, Scott set up base at Cape Evans, about 21 kilometres (13 miles) north. A living hut was divided into two sections, following the Royal Navy tradition of segregating officers and "lower deck". Soon thereafter, parties began to head out, with one, under Scott, going south onto the Great Ice Barrier to lay depots for the trek on the Pole; one going on a geological survey of the western side of McMurdo Sound and one sailing in *Terra Nova* towards King Edward VII Land on the eastern side of the Barrier. This Eastern Party, under Victor Campbell, was amazed to find Amundsen securely ensconced at Framheim, his base at the Bay of Whales on the Barrier (see page 48). Not finding anywhere else to land, they returned to Cape Evans to report about Amundsen and were then taken to Cape Adare, thus becoming the Northern Party.

Scott, meanwhile, had less success than hoped, and after a variety of problems left the bulk of his supplies at One Ton Depot, located at 79°29′ S, some 57 kilometres (or 31 minutes of latitude) short of where he had hoped to locate it. He and his party then returned to Cape Evans for the winter, during which his old friend Edward Wilson oversaw the scientific programme.

ABOVE The two sides of a medal issued by the RGS honouring the achievements of Scott's last expedition. The obverse is an image of Captain Scott, and the back shows the five members of the polar party.

BELOW The ponies in the stables. Oates treated them with constant care and devotion, even though he considered them a poorer quality than he would have chosen himself.

LEFT Four of the five men of the final polar party, sledging across the Plateau. Bowers, who was without skis, took this picture of, from left: Evans, Oates, Wilson and Scott.

The dog teams and a number of support personnel were sent back from the base of the Beardmore, and 12 men continued on, having been divided into three man-hauling teams. The first of these was sent back from a depot near the top of the 200-kilometre-long (125-mile) river of ice on 21 December. Two weeks later, Scott sent back another party under his second-in-command, Teddy Evans. But from that group, he kept one man, Henry "Birdie" Bowers, an act that threw out of kilter the carefully planned food and fuel allowances.

The final party – Scott, Bowers, Wilson, L E G Oates and Edgar Evans – slogged across the Polar Plateau, finally reaching the South Pole on 17 January 1912 to discover Amundsen had preceded them by about a month (see page 49). Their return was plagued by extreme disappointment, lack of food and fuel, exceptionally low temperatures and scurvy. Evans broke down on the descent of the Beardmore, and died near its base on 17 February. In March, Oates, so debilitated by frostbite, scurvy and possibly gangrene that continuing would condemn the others to failure, walked out of the tent to his death. The other three struggled on, but their food and fuel dwindled away, and in late March they died in their tent, only 18 kilometres (11 miles) short of re-supply at One Ton Depot.

The next spring, a search party found the bodies of Scott, Bowers and Wilson, along with their diaries and correspondence. They had lost the race to the Pole, but their tragic and noble deaths earned Scott a place as one of Britain's greatest heroes of exploration.

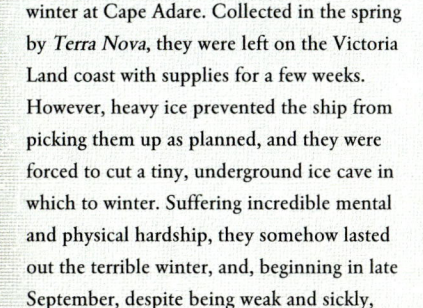

ATTACHMENT
Scott's diary entry for 16 March 1912, telling the now-famous story about how Captain Oates left the tent to go to his death in the hopes that it would mean his comrades – Scott, Wilson and Bowers – might live. His plea in his final entry, dated 29 March 1912, for the explorers' families to be looked after, received an overwhelming response from the generous British public.

BELOW The five most disappointed men in the world. After all of their struggles, they had found that Amundsen and his four companions had beaten them to the last place on Earth – the South Pole. From left: Wilson, Scott, Evans, Oates and Bowers.

Scott's planned assault on the South Pole involved four types of motive power: ponies, dogs, motor sledges and man-hauling. However, from early on there were problems, as several ponies were lost on the depot-laying journey and one motor sledge went through the ice and sank when it was being offloaded. When, on 1 November, Scott and nine others left Cape Evans with 10 ponies, a number of the scientists were included as support personnel. It was not long until Scott found that the two remaining motor sledges – which had left several days before the pony party – had broken down. The ponies that did survive the brutal march across the Barrier were killed near the bottom of the Beardmore Glacier.

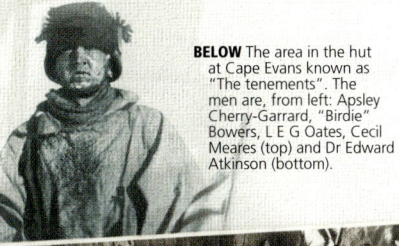

BELOW The area in the hut at Cape Evans known as "The tenements". The men are, from left: Apsley Cherry-Garrard, "Birdie" Bowers, L E G Oates, Cecil Meares (top) and Dr Edward Atkinson (bottom).

THE NORTHERN PARTY

After their change of destination, the six members of the Northern Party spent the winter at Cape Adare. Collected in the spring by *Terra Nova*, they were left on the Victoria Land coast with supplies for a few weeks. However, heavy ice prevented the ship from picking them up as planned, and they were forced to cut a tiny, underground ice cave in which to winter. Suffering incredible mental and physical hardship, they somehow lasted out the terrible winter, and, beginning in late September, despite being weak and sickly, made a remarkable 37-day march to safety at Cape Evans.

BELOW The Northern Party after their terrible winter. From left: George Abbott, Victor Campbell, Harry Dickason, Raymond Priestley, Murray Levick and Frank Browning.

HELL IN MID-WINTER

Before the southern journey, Wilson and Bowers had already endured the harshest Antarctic conditions imaginable. Since the *Discovery* Expedition, Wilson had hoped to study eggs collected during emperor penguins' winter incubation. In mid-winter 1911, he, Bowers and Apsley Cherry-Garrard made a trek to the Cape Crozier rookery. Relaying sledges totalling 343 kilograms (757 pounds) at temperatures dropping to -61° C (-77° F), they took more than two weeks to cover the 105 kilometres (65 miles). Once there, gale-force winds carried away their tent, which, remarkably, they later found. With this reprieve, they somehow managed to trek back to Cape Evans.

ANTARCTIC 1910–1912
AMUNDSEN ATTAINS THE POLE

ROALD AMUNDSEN HAD A SPIRIT TOO RESTLESS TO BE SATISFIED WITH SIMPLY BEING THE FIRST MAN EVER TO NAVIGATE THE NORTHWEST PASSAGE (SEE PAGES 40–41), SO NOT LONG AFTER HIS RETURN TO NORWAY IN 1906, HE TURNED HIS SIGHTS ON A NEW GOAL: THE NORTH POLE. HIS PLAN INCORPORATED A DRIFT IN THE ARCTIC OCEAN SIMILAR TO FRIDTJOF NANSEN'S, AND HE SUCCEEDED IN SECURING *FRAM* FOR THE EXPEDITION. BUT AMUNDSEN'S PLANS WERE DESTROYED IN SEPTEMBER 1909 WHEN, FIRST HIS OLD COLLEAGUE FREDERICK COOK AND THEN ROBERT E PEARY, CLAIMED TO HAVE REACHED THE POLE.

ABOVE *Fram* at the Bay of Whales. In the background is Scott's ship *Terra Nova*, the crew of which was shocked – and dismayed – to find the Norwegians building their base right onto the Great Ice Barrier.

BELOW A winter evening at Framheim. Although the men worked constantly to improve their sledging gear, train the dogs and prepare in every way possible, the design of the base made it extremely pleasant.

Amundsen quickly decided upon a new goal – the South Pole – but, worried about losing both his funding and the use of *Fram*, he decided not to let anybody know about his *volte face* until it was too late to stop him. While his meticulous preparations went ahead, he disappeared from public view, not wishing to lie outright, particularly to Nansen or Robert Falcon Scott, who was attempting to consult him about his own British expedition. Even the men Amundsen hired for his shore party did not know his plans until September 1910, when *Fram* reached Madeira. But given a choice, every man decided to follow their leader.

In January 1911, *Fram* arrived at the Bay of Whales, along the Great Ice Barrier, on which Amundsen's party established their base, Framheim. Shortly thereafter, Scott's Eastern Party encountered them. Before the Sun disappeared for the winter, Amundsen's sledging party – men expert in skiing and driving dogs – set up depots as far south as 82°.

Throughout the winter, the men worked to reduce excess weight in their equipment; prepared and packed their sledging rations and made every effort to perfect their clothing and other gear. At the same time, each was responsible for a number of the dogs. But through it all, Amundsen fretted that Scott – with his motor sledges – would beat him to the Pole. These worries led Amundsen into committing his greatest error.

In early September, despite the Sun having only recently returned, Amundsen and eight men headed south, leaving cook Adolf Lindstrøm alone at the base. Within days, the temperature plummeted to −56°C (−69°F), and by the time they reached the first depot, several men were suffering from frostbite and several dogs had frozen to death. The party retreated to Framheim, where Amundsen changed his plans, limiting the polar party to five men.

In mid October Amundsen was on the move again, leaving Framheim with four companions: Helmer Hanssen,

ABOVE Digging the foundations of Framheim. Establishing a base on the Great Ice Barrier provided many challenges, but the result was that the Norwegians started a full degree closer to the Pole than Scott's party.

ENCLOSURE Pages from Amundsen's diary during his party's approach to and arrival at the South Pole. On the day they attained their goal, his four comrades made sure Amundsen was in the lead, in order that he be unquestionably the first man to reach the Pole (see translation page 60).

BELOW A letter written by Amundsen to King Haakon VII of Norway at the South Pole. Amundsen left it in the tent he named Polheim, where it was collected by Scott's party. It was later found with the bodies of Scott and his companions.

who had been on *Gjøa* with him; Sverre Hassel, who had accompanied Otto Sverdrup on the second Fram expedition; champion skier Olav Bjaaland and Oscar Wisting. They headed straight across the Barrier, taking with them 52 dogs. They soon were in totally new terrain, and south of 85° they faced the Transantarctic Mountains. These they overcame on a remarkable ascent to the Polar Plateau, conquering the previously unknown Axel Heiberg Glacier. With Amundsen's careful planning and his men's extensive background in such conditions, they took every obstacle in their stride while averaging 24 kilometres (15 miles) a day in five to six hours of travel – which left large amounts of time each day for men and dogs to recover.

The result was that, on 7 December 1911, Amundsen and his party passed Shackleton's farthest south, and a week later they reached the vicinity of the South Pole, more than a month ahead of Scott. They spent several days making observations and skiing in each direction to make certain that they had attained the Pole regardless of the variability of their measuring equipment. They raised a tent, named it Polheim, left a message for Scott, and turned back north.

With depots every degree all the way back, the five men made the return journey with relative ease. In late January 1912, they arrived at their base to find *Fram* already waiting. All that was left to do was to transfer the equipment and materials from Framheim to the ship, and to sail north to announce their triumph.

ABOVE Part of a sextant that was left by Amundsen at Polheim. As with the letter, this was found in Scott's final tent, proving conclusively that the Norwegians had beaten the British to the South Pole.

RIGHT The Norwegians in the vicinity of the South Pole on 14 December 1911. Helmer Hanssen took the photograph of (from left): Oscar Wisting, Olav Bjaaland, Sverre Hassel and Amundsen.

LEFT Emotionally overcome by being dismissed from first the polar party and later the expedition, in January 1913 Johansen committed suicide.

TROUBLES WITH JOHANSEN

Among Amundsen's shore party, there was one man he had not willingly selected. Hjalmar Johansen had been Nansen's companion on his farthest north (see pages 22–23), and when Nansen asked Amundsen to take Johansen, he felt obliged to do so. But there was unease between them, as Johansen was the only man with experience to rival Amundsen's. After Amundsen's abortive September start resulted in several men being frostbitten, Johansen exploded at Amundsen in front of the others. This led Amundsen to decide the polar party would consist of only five men, while Johansen was relegated to a trip towards King Edward VII Land.

SHIRASE'S JAPANESE EXPEDITION

Scott's party was not the only one to visit Framheim. A year later, a Japanese expedition under Nobu Shirase spied *Fram* waiting at the edge of the Great Ice Barrier. Shirase originally left Japan in December 1910, but could not penetrate deep into the Ross Sea owing to heavy ice. The expedition wintered in Sydney, where lack of funds made life extremely difficult until T W Edgeworth David interceded on their behalf. In January 1912, Shirase was back at the Barrier, and led a "Dash Patrol" 237 kilometres (147 miles) south, while another party investigated King Edward VII Land.

ABOVE Shirase's "Dash Patrol" at latitude 80°5'S. Upon arrival at the Great Ice Barrier, the inexperienced members of the Japanese expedition saw *Fram* and were worried she was a pirate ship.

49

ANTARCTIC 1911–1913
FILCHNER FOILED BY THE ICE

WILHELM FILCHNER WAS ALREADY A WELL KNOWN TRAVELLER WHEN HE FIRST PROPOSED A CROSSING OF ANTARCTICA IN ORDER TO DETERMINE IF IT WERE A SINGLE CONTINENT OR A SERIES OF SEPARATE ISLANDS. FILCHNER'S ORIGINAL PLAN WAS TO HAVE TWO SHIPS, ONE GOING TO THE ROSS SEA AND ONE TO THE WEDDELL SEA, SETTING DOWN PARTIES THAT WOULD MEET MID-CONTINENT. HOWEVER, LACK OF FUNDING EVENTUALLY PRECLUDED A SECOND VESSEL.

The Norwegian ship *Björnen*, which had been specifically built for operating in the polar seas, was purchased for the expedition, refitted and renamed *Deutschland*. Under Captain Richard Vahsel, she left Bremerhaven in May 1911 and reached South Georgia in October. Seven weeks later, after a brief visit to the South Sandwich Islands, the party entered the eastern Weddell Sea, where in January 1912 they discovered the Luitpold Coast.

In February, an anchorage was found at Vahsel Bay, an inlet in a vast ice shelf at the base of the Weddell Sea. Filchner named the ice shelf itself after Kaiser Wilhelm, but the emperor later changed it to the Filchner Ice Shelf. In the following days, the expedition's ponies, dogs and supplies were unloaded and a hut measuring 17.5 by 9 metres (57 by 30 feet) was built for the party to live and work in. But early one morning, as it neared completion, deafening reports from the ice announced that a vast section of the ice shelf had suddenly calved. The huge iceberg on which the hut stood began to float away to the north. Filchner later determined that the event was caused by a springtide measuring up to three metres (10 feet) coupled with a sudden falling of the barometric pressure. The area it affected was approximately 600 square kilometres (230 square miles).

As the iceberg drifted towards the open sea, Filchner's party spent two feverish days dismantling the hut and taking it and the animals back to the ship,

ABOVE On 26 March 1912, a meteorological balloon was sent up with recording instruments to take measurements in the upper atmosphere. However, it soon had to be hauled down as a result of heavy winds.

BELOW Filchner and Kling search the horizon for Morrell's Land, or New South Greenland, which had been reported in 1823. Struggling through terrible winter conditions, they showed no such land existed.

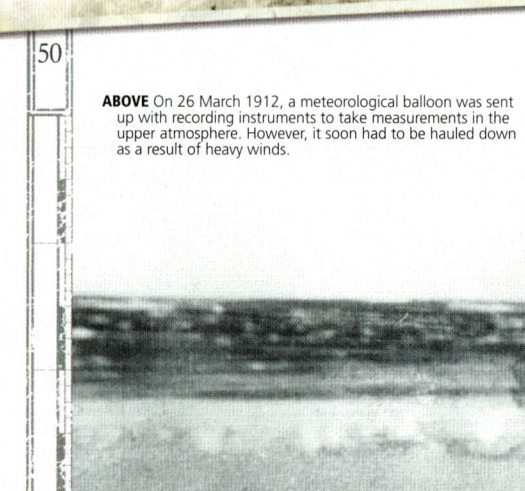

which remained near to the berg. However, large amounts of broken ice followed them north and prevented a subsequent return to the ice shelf. In early March, after a brief landing was made on the continent itself and depots established, the sea froze over extremely rapidly, and *Deutschland* was bound tightly into the drifting pack ice.

For the next nine months *Deutschland* drifted northwest. To make the best of the situation, the party built stables for the expedition's ponies, put up kennels for the dogs and erected small huts and tents, which housed scientific equipment. Owing to Filchner's foresight, the auxiliary boiler aboard ship could be fuelled by the carcasses of penguins and the blubber of seals, allowing the men to preserve the ship's coal.

In June 1912, in the midst of winter, Filchner, Alfred Kling and Felix König sledged across the sea ice towards the position of "Morrell's Land", which the American sealer Benjamin Morrell had claimed he had seen in 1823. Working in daylight that lasted only two to three hours, in temperatures as low as -35°C (-31°F), the three reached the position where the island should have been located, but found by measurements through the ice that they were still in an area of very deep water, indicating that Morrell's Land did not actually exist. The return journey to the ship was extremely difficult, as wide leads had opened in the ice, which had also thinned dangerously in many other places. In addition, *Deutschland* had drifted 61 kilometres (38 miles) from its previous position, and only remarkable navigational efforts by Kling allowed them to find the ship.

In August, as the winter came to a close, Captain Vahsel died of a long-term illness. Wilhelm Lorenz, the first officer, thereupon took command of *Deutschland*, although he was replaced as captain by Kling before the expedition reached Europe. Finally, in November, the ship broke free of the ice. In mid-December, the expedition reached South Georgia, from where they headed north.

The next year, *Deutschland* was purchased for an Austrian expedition under König. Renamed *Osterreich*, in August 1914 she was ready to sail from Trieste when the First World War broke out. The expedition never sailed.

ABOVE In April, during the drift, Filchner had three individual huts built for recording various scientific measurements. The geodetic hut, shown here, later had the balloon hut built right onto it.

- Filchner, 1911–1912
- Filchner, beset in ice
- Shackleton, 1914–1916 (see pages 54–55)
- Shackleton, beset in ice (see pages 54–55)
- Shackleton in *James Caird* (see pages 54–55)
- Ellsworth, 1935 (see pages 56–57)

RIGHT The precise fate of *St Anna*, shown here before her departure from Alexandrovsk (now Murmansk), will never be known.

RIGHT Wilhelm Filchner in 1920, when he was trying to raise money for his self-funded expedition that began in 1925.

Filchner's Asian Expeditions

Filchner spent most of his career in the mountains of central Asia, exploring and conducting geomagnetic research. In 1900, at only 22, he travelled there on his own, but – caught in the politics of the "Great Game" between Britain and Russia – was forced to leave. His travel narrative attracted official interest, and in 1903–1905 he was placed in charge of a geomagnetic and topographical survey of Tibet, where he returned during the period from 1925 to 1928. After going back yet again in 1935 for geomagnetic work, he was reported as missing, but turned up in Kashgar in western China in 1937. In 1939–1940, he conducted similar studies in Nepal.

Caught in the Ice a World Away

While Filchner was caught in the Antarctic ice, Georgi Brusilov's Russian expedition was trapped just as mercilessly in the Arctic. Within a few months of Brusilov's departure in July 1912, his ship, *St Anna*, was snared by the ice of the Kara Sea and started drifting north. In April 1914, the navigator, Valerian Albanov, led 14 men south from the ship; those left aboard were never seen again. One by one, those in Albanov's party died, and only he and Aleksandr Kondrat managed to reach Cape Flora on Franz Josef Land, where they were rescued by another Russian ship.

ANTARCTIC 1911–1914: MAWSON'S RACE WITH DEATH

NOT LONG AFTER RETURNING FROM SHACKLETON'S BRITISH ANTARCTIC EXPEDITION, ON WHICH HE HAD BECOME THE DE FACTO LEADER OF THE PARTY ATTEMPTING TO REACH THE SOUTH MAGNETIC POLE (SEE PAGES 42–43), DOUGLAS MAWSON INITIATED THE IDEA OF LEADING HIS OWN EXPEDITION TO INVESTIGATE THE VAST UNKNOWN AREA SOUTH OF AUSTRALIA. SCOTT THEREUPON OFFERED MAWSON A PLACE ON HIS FINAL POLAR PARTY (WHICH WOULD CONSIST OF ABOUT FOUR PEOPLE) IF HE ACCOMPANIED THE EXPEDITION, BUT MAWSON DECLINED. INSTEAD, HE BEGAN PLANNING A SERIOUS SCIENTIFIC RESEARCH EXPEDITION IN CONCERT WITH SHACKLETON.

In late 1910, Shackleton withdrew from participation in the expedition, so Mawson moved ahead alone with the most extensive Antarctic effort yet planned: the Australasian Antarctic Expedition (AAE). Mawson's ultimate intention was to man four bases, three on the continent and one at sub-Antarctic Macquarie Island. The shore parties at these bases would carry out a wide range of scientific studies as well as geographical exploration, while the ship-borne personnel would make multiple sub-Antarctic voyages to obtain oceanographic and biological data.

The expedition sailed from Hobart, Tasmania, in December 1911 on the converted whaler *Aurora*, commanded by John King Davis. Unable to find suitable landing sites after leaving one party at Macquarie Island, Mawson consolidated his three Antarctic bases into two. He personally oversaw Main Base at Cape Denison in Commonwealth Bay, with a total of 18 men. The Western Base, with a party of eight commanded by Frank Wild, was located some 2,400 kilometres (1,500 miles) to the west on the Shackleton Ice Shelf. Unbeknownst to Mawson when Main Base was established in January 1912, Cape Denison is the windiest place on Earth. The men were to find that the winds, which averaged more than 80 kilometres (50 miles) per hour for seven consecutive months and reached a high of more than 320 kilometres (200 miles) per hour, affected everything done there.

The AAE included the largest number of scientists that had yet been to the Antarctic. In early summer 1912 Mawson also sent out five major sledging parties from Cape Denison to take scientific measurements and compile in-depth information about the area. One party, commanded by Robert Bage and including photographer Frank Hurley and magnetician Eric Webb, headed towards the South Magnetic Pole, approaching it from the opposite direction from which Mawson had several years before. Another party, under Cecil Madigan, explored much of the coastline to the east, not turning back until they had travelled 435 kilometres (270 miles) and crossed two magnificent glacier tongues.

Mawson's own party went the farthest, using dogs rather than engaging in man-hauling as the other groups did. In mid-December 1912, at a point about 500 kilometres (310 miles) from Cape Denison, one of Mawson's two companions, Belgrave Ninnis, was lost down a crevasse with most of the food and half the remaining dogs. Mawson and his other colleague, Xavier Mertz, quickly turned back to base, hoping to reach it before their supplies failed. But both soon began suffering the debilitating affects of

ABOVE Douglas Mawson envisioned an expedition that would concentrate on science rather than adventure. Nevertheless, his own sledging journey turned out to be one of the greatest survival stories in history.

ABOVE Frank Hurley first became famous for his magnificent photographs from the AAE. This one shows the expedition ship *Aurora* framed by an iceberg while anchored off of Cape Denison in 1912.

ABOVE Hurley took this evocative photograph of Leslie Whetter and John Close quarrying ice in the midst of a blizzard at Cape Denison. The ice was collected no matter what the conditions, and melted for water in the hut.

hypervitaminosis A, contracted by eating the livers of the dogs. When Mertz died from this condition on 8 January 1913, Mawson was left to continue alone, in a very sick condition. In perhaps the greatest feat in Antarctic history, he struggled on for a month, somehow managing to reach Cape Denison to find that *Aurora*, which had been waiting his return for four weeks, had left hours before in order to pick up Wild's party.

Six men had remained behind to search for Mawson, and this much smaller party remained at Cape Denison for another year. But while Mawson managed to recover from his ordeal, the wireless operator, Sydney Jeffryes, went mad, leaving the other men worried, frightened and mystified as to how to help him. Finally, in December 1913, Davis brought *Aurora* into Commonwealth Bay for the third time. The shore party boarded the ship, and Mawson and company travelled west to where Wild's party had been based. Only after further investigation of the Shackleton Ice Shelf and the surrounding area, did they finally turn north for Australia.

RIGHT With snow covering the hut at Western Base, the men built a series of underground storage grottoes, and a long tunnel to reach the outside.

ABOVE Xavier Mertz comes out of the tunnel to Aladdin's Cave. This was a room dug into the ice near the top of the long, incredibly steep slope leading up to the Plateau. It served as a respite from the brutal local winds.

ATTACHMENT
Mawson's diary from the days preceding the death of Mertz, when, despite being ill himself, he devoted most of his time to caring for his friend. Mertz's inability to travel during this period, while still using their meagre food supplies, made Mawson's subsequent solo trek even more difficult.

WILD'S WESTERN BASE

After weeks sailing west searching for a landing site, the eight-man Western Party finally settled on a location not even on solid land: the edge of the Shackleton Ice Shelf. In the following year, after several depot-laying efforts and an abortive strike straight inland, Wild led one party to the eastern side of the ice shelf, discovering one of the world's most powerful rivers of ice, the magnificent Denman Glacier. Meanwhile, Sydney Jones and two others headed west, reaching Gaussberg, thereby linking the Australians' discoveries to the earlier ones of Drygalski's German expedition (1901–1903) (see pages 36–37)

JOHN KING DAVIS

During the Australasian Antarctic Expedition, Davis began to establish himself as the greatest of all Antarctic sea captains. Mawson had met Davis on Shackleton's British Antarctic Expedition, when he had been first mate of *Nimrod*. As master of *Aurora* and second-in-command of the AAE, Davis was in charge of all three voyages south, of the oceanographic research programme, and of raising funds for the unexpected third trip to the Antarctic. Davis later commanded *Aurora* for the rescue of the Ross Sea Party on Shackleton's Imperial Trans-Antarctic Expedition, and spent many years as the Commonwealth Director of Navigation.

Map legend:
- Magnetic Pole journey, 1909 (see pages 42–43)
- Cruise for AAE, 1911–1912
- Mawson's Far-Eastern Party, 1912–1913
- East Coast Party, 1912–1913
- Southern Party, 1912–1913
- BANZARE, 1929–1930 (see pages 58–59)
- BANZARE, 1930–1931 (see pages 58–59)

ANTARCTIC 1914–1917

SHACKLETON'S ENDURANCE

BELOW A team of sledge dogs looks on as the ice continues slowly to crush the doomed remains of *Endurance*. The dogs didn't last much longer once the ship went down, as they were killed to preserve food.

BELOW The photographer Frank Hurley and Shackleton in front of the tent they shared at Patience Camp. Hurley is skinning a penguin as fuel for the blubber stove sitting between them, which he built.

WITH ROALD AMUNDSEN AND ROBERT FALCON SCOTT BOTH REACHING THE SOUTH POLE, ERNEST SHACKLETON NEEDED A NEW QUEST IN HIS PURSUIT OF FAME AND FORTUNE. THE VENTURE HE SELECTED HAD INITIALLY BEEN PROPOSED BY W S BRUCE AND LATER ATTEMPTED BY WILHELM FILCHNER: THE CROSSING OF THE ANTARCTIC CONTINENT. BUT NOW IT WAS REBORN AS THE IMPERIAL TRANS-ANTARCTIC EXPEDITION (ITAE).

Shackleton's plan was for his main party to land at Vahsel Bay at the base of the Weddell Sea. Six men using dogs would head for the South Pole, and thence on to McMurdo Sound, following Shackleton's earlier route (see pages 42–43). A second party would establish a base on Ross Island, then set out depots across the Great Ice Barrier to the Beardmore Glacier, so that the main party could be re-supplied on the way.

When Shackleton announced his expedition, nearly 5,000 men (and three schoolgirls) applied to join it. His party was reduced, however, just as it was about to sail, when a number left for military duty after war was declared on 5 August 1914. Patriotically, Shackleton offered his ships – *Endurance*, intended for the Weddell Sea, and *Aurora*, which he had purchased from Douglas Mawson, for the Ross Sea – to the British government. But in a telegram, the First Lord of the Admiralty, Winston Churchill, simply replied: "Proceed". By late October, *Endurance* had reached South Georgia, and in December, after a series of repairs, she struck out for the Weddell Sea. Coates Land was sighted on 10 January 1915, but with Vahsel Bay not far away, the sea froze over and *Endurance* was soon trapped.

For the following nine months they drifted lazily northwards, and in October, after the ship had begun to leak owing to ice pressure, Shackleton established a base nearby on the ice. After *Endurance* was crushed and sank in November, the company attempted a march across the floes to improve their chances of reaching land, but six brutal days took them only 16 kilometres (10 miles), so they settled in at what was called Patience Camp to wait for the drift to bring them to the open sea. Meanwhile, the dogs were killed to preserve the food supplies.

Shackleton considered trying to reach the South Orkney Islands, the Antarctic Peninsula or Paulet Island, where a hut and supplies remained from Otto Nordenskjöld's expedition (see pages 34–35). But by April 1916, when their ice camp finally reached the open sea, the best option appeared to be Elephant Island. The men took to open boats and after sailing and rowing for seven days made a precarious landing there.

Shackleton now decided to go for help, and on 24 April, leaving Frank Wild, his second-in-command, in charge,

ENCLOSURE

1. A letter from Shackleton to J Scott Keltie, the secretary of the RGS. Shackleton claimed that the War Office was solidly backing his new expedition, and expected the Admiralty to soon follow suit.

2. An extremely long letter from Sir Clements Markham, formerly the president of the RGS, to the Society's Council, in which he denigrated Shackleton's Antarctic plans and experience in the hopes that the Council not give recognition to his upcoming expedition.

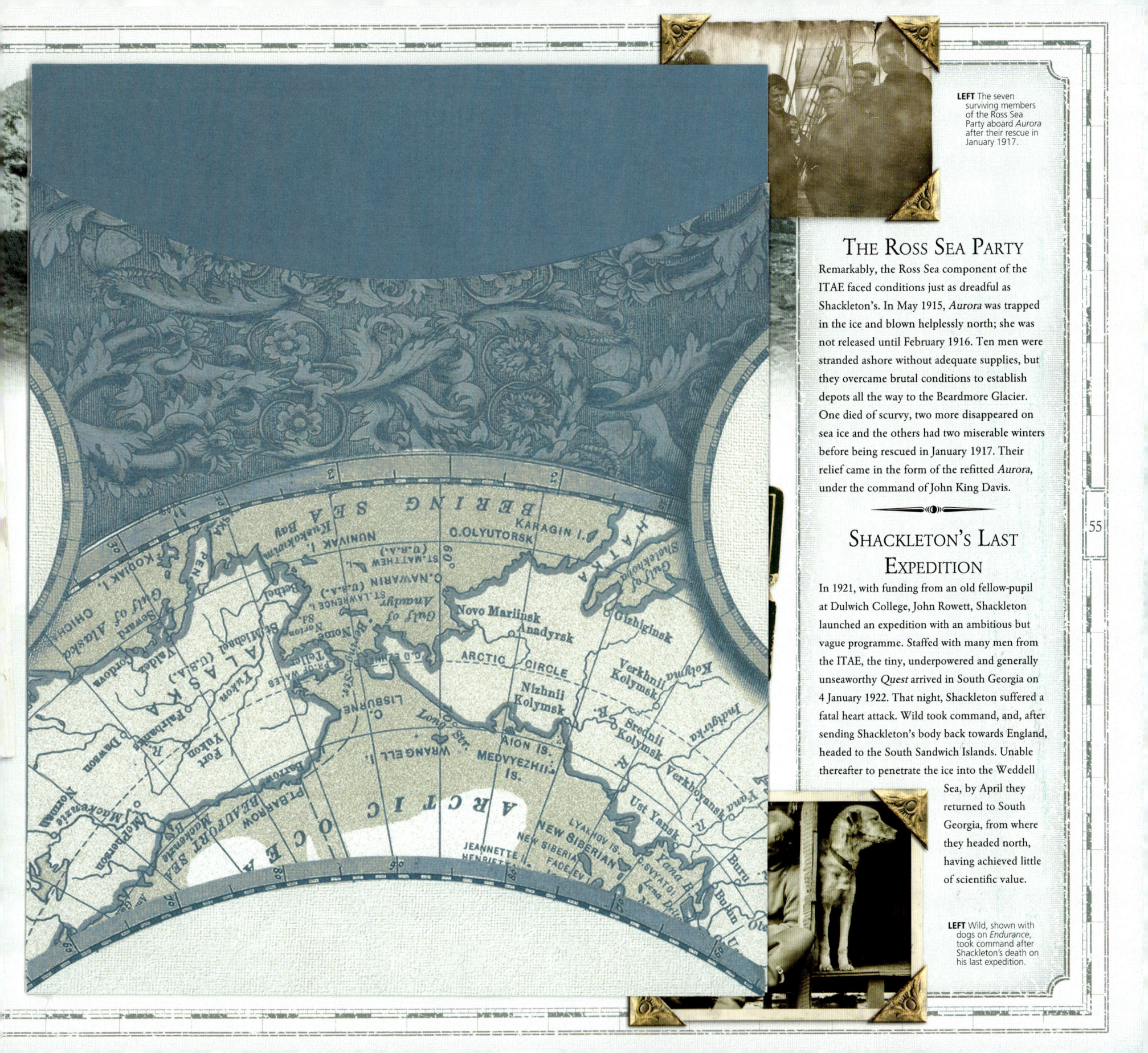

LEFT The seven surviving members of the Ross Sea Party aboard *Aurora* after their rescue in January 1917.

THE ROSS SEA PARTY

Remarkably, the Ross Sea component of the ITAE faced conditions just as dreadful as Shackleton's. In May 1915, *Aurora* was trapped in the ice and blown helplessly north; she was not released until February 1916. Ten men were stranded ashore without adequate supplies, but they overcame brutal conditions to establish depots all the way to the Beardmore Glacier. One died of scurvy, two more disappeared on sea ice and the others had two miserable winters before being rescued in January 1917. Their relief came in the form of the refitted *Aurora*, under the command of John King Davis.

SHACKLETON'S LAST EXPEDITION

In 1921, with funding from an old fellow-pupil at Dulwich College, John Rowett, Shackleton launched an expedition with an ambitious but vague programme. Staffed with many men from the ITAE, the tiny, underpowered and generally unseaworthy *Quest* arrived in South Georgia on 4 January 1922. That night, Shackleton suffered a fatal heart attack. Wild took command, and, after sending Shackleton's body back towards England, headed to the South Sandwich Islands. Unable thereafter to penetrate the ice into the Weddell Sea, by April they returned to South Georgia, from where they headed north, having achieved little of scientific value.

LEFT Wild, shown with dogs on *Endurance*, took command after Shackleton's death on his last expedition.

POLAR 1922–1935

THE AGE OF FLIGHT

THE 1920S SAW THE BEGINNING OF A NEW ERA IN POLAR EXPLORATION – ONE MARKED BY NEW TECHNOLOGY, AND PARTICULARLY BY THE USE OF AEROPLANES AND AIRSHIPS. SUCH EFFORTS WERE NOT TOTALLY NEW, OF COURSE. SALOMON ANDRÉE ATTEMPTED TO FLY TO THE NORTH POLE IN A BALLOON IN 1897, AND IN THE FIRST DECADE OF THE TWENTIETH CENTURY WALTER WELLMAN MADE THREE ATTEMPTS IN A DIRIGIBLE (SEE PAGES 24–25). DOUGLAS MAWSON PLANNED ON TAKING AN AEROPLANE TO THE ANTARCTIC, BUT IT CRASHED AT A FUND-RAISING EVENT IN ADELAIDE. BUT IN THE YEARS FOLLOWING THE FIRST WORLD WAR, FLIGHT BECAME PROGRESSIVELY MORE COMMON, AND MORE SUCCESSFUL.

One of the leaders in the use of aircraft in the Arctic was the former master of dog travel: Roald Amundsen. After several relatively unsuccessful flight attempts in 1922 and 1923, Amundsen was approached by Lincoln Ellsworth, an American who was enthusiastic not only about flight, but about working with the great explorer. In 1925, they attempted to fly two Dornier flying-boats from King's Bay in Spitsbergen to the North Pole. Both aeroplanes developed problems en route, and were forced to land on the ice at 87°44′ N. After a month of making repairs and hewing a runway, the six expedition members piled into one plane and managed to get aloft to return to Svalbard.

The next year, Amundsen and Ellsworth unveiled a different plan of attack: an airship. Ellsworth again provided the money and Amundsen the charismatic leadership, and the designer and pilot was the Italian Umberto Nobile. The party was ensconced at King's Bay in late April 1926 when some 50 members of an American expedition headed by Richard E Byrd arrived. With a Fokker tri-motor named *Josephine Ford*, Byrd was planning on flying to the Pole. Downplaying any rivalry, Amundsen allowed Byrd to do so first. On 9 May, Byrd flew off, and upon his return 13½ hours later, declared that he had reached the Pole – a claim widely dismissed after his flight diary was discovered seven decades later and cast doubts upon his statements. Two days after Byrd returned, Amundsen and his party headed north in the dirigible *Norge*, and, after a flight of 16 hours, they reached the Pole. They then continued on to Teller, Alaska, completing the first crossing by air of the Arctic Basin.

The fame Byrd gained from his claim allowed him to organize and lead a series of expeditions to the Antarctic. On the first of these, he carried out extensive aerial reconnaissance, highlighted, on 28–29 November 1929, by

TOP LEFT Ellsworth and Amundsen after their 1925 flight towards the North Pole. Their sudden return after being missing meant the authorities were initially unconcerned when Amundsen vanished three years later.

LEFT Byrd in his flight gear. His claim to have reached the North Pole has long been disputed, but his perceived success endowed him with a celebrity status that allowed him to organize his Antarctic expeditions.

ENCLOSUREs

1. Pages from *Little America Times*, a newsletter published in the US (1933–1935) to report about Byrd's second expedition and, in this case, Ellsworth's attempted Antarctic flight. It was not until his next attempt that Ellsworth successfully flew across the Antarctic.

2. The orders from Admiral Chester Nimitz recalling Byrd to active duty. Byrd was subsequently placed in command of the United States Antarctic Service Expedition (1939–1941), which was his third Antarctic command.

flying to the South Pole from his base at Little America on the Ross Ice Shelf. Byrd's flights were not, however, the first in the Antarctic. These were made the previous year on an expedition under Hubert Wilkins, who had earlier served on Shackleton's last expedition and then established a reputation for aerial work in the Arctic. In November 1928, Ben Eielson made the first Antarctic test flight at Deception Island, and the next month he and Wilkins flew the first significant distance, crossing the Antarctic Peninsula. The next summer, Wilkins made more flights over little-known parts of the Peninsula region.

In the following years, Ellsworth also turned from the far north to the Antarctic, but his goal was to cross the continent from the Peninsula to the Ross Sea. In 1933 and again in 1934, he launched expeditions, but was prevented by conditions from conducting extensive flight operations. Finally, on his third attempt, on 23 November 1935, Ellsworth left Dundee Island with Herbert Hollick-Kenyon. They projected the flight would take about 14 hours, but refuelling and bad weather forced them to land for extended periods four times, and they did not reach the vicinity of Little America until mid-December, when they ran out of fuel some 26 kilometres (16 miles) short. The two then walked to Byrd's former base, where they stayed comfortably until they were collected by a search-and-rescue party.

ABOVE RIGHT Ellsworth's plane *Polar Star* is lifted aboard *Wyatt Earp*. Heavy seas broke up the sea ice that the plane was on after a test flight, and she sank into the sea, suffering serious damage despite being quickly retrieved.

ABOVE Work being conducted on Byrd's tri-motor Fokker, *Josephine Ford*. The plane was named for the daughter of Edsel Ford, the head of the automobile dynasty, who gave financial support to the expedition.

ABOVE Byrd (left) is congratulated by US President Franklin Roosevelt in May 1935. Byrd returned to the US as a hero after his second Antarctic expedition, having twice just barely survived while wintering by himself at a small weather station.

- Andree, 1897 (see pages 24–25)
- Amundsen & Ellsworth, 1925
- Amundsen & Ellsworth, 1926
- Nobile, 1928
- Kuznetsov, 1948

RIGHT Russian rescuers inspect Einar Lundborg's overturned Fokker, which crashed on his second attempt to land at Nobile's camp on the ice.

THE *ITALIA* DISASTER

Following the crossing of the Arctic Basin by *Norge*, Nobile felt that his role had been unfairly overshadowed by Amundsen and Ellsworth. So two years later, in May 1928, he flew a new airship, *Italia*, to the North Pole. However, it crashed on its return to Spitsbergen, leaving nine men stranded on the ice; the others were never seen again. An international search-and-rescue effort followed, during which Amundsen himself vanished. In June, the Italians were finally located by a Swedish pilot, who brought out the injured Nobile before his plane crashed on a subsequent attempt landing on the ice. The next month, the Soviet icebreaker *Krassin* rescued the remaining men.

THE RUSSIANS AT THE NORTH POLE

If, as is generally believed, the flight of *Norge* was the first attainment of the North Pole, when did anyone actually indisputably stand there? The answer is 23 April 1948, when three planes on a Soviet secret mission, led by Aleksandr Kuznetsov, landed there with 24 men, including 15 crew, four scientists and two journalists. It has been stated that the first man on the ice was hydrologist Pavel Gordiyenko, but, according to magnetician Pavel Senko, the last surviving member of the expedition, with the men piling out of the planes no-one knew certain.

ABOVE *Norge* in a vast hangar during her journey north from Italy to Norway to Spitsbergen. Her flight to King's Bay took just under four weeks.

POLAR 1929–present

AFTER THE HEROIC AGE

BELOW A Dornier Super Wal hydroplane sits on the steam-driven catapult used to launch the planes from *Schwabenland* during the German expedition of 1938–1939. A vast area was surveyed and claimed for Germany.

SINCE THE END OF THE HEROIC AGE, GEOGRAPHICAL EXPLORATION IN THE POLAR REGIONS HAS GIVEN WAY TO SCIENTIFIC RESEARCH, POLITICAL CONSIDERATIONS AND, MORE RECENTLY, ADVENTURE. FOR DECADES AFTER ROALD AMUNDSEN CROSSED THE ARCTIC BASIN AND RICHARD E BYRD CLAIMED THE NORTH POLE, THE PUBLIC TENDED TO THINK INFREQUENTLY ABOUT THE FAR NORTHERN REGIONS, ALTHOUGH THE 1920S AND 1930S SAW A SERIES OF UNIVERSITY EXPEDITIONS TO THE ARCTIC FROM BOTH OXFORD AND CAMBRIDGE.

ABOVE TOP A cheque addressed to the well known bookseller Francis Edwards on behalf of the Norwegian-British-Swedish Antarctic Expedition. Not only was the expedition supplied with practical texts, it also carried works by classic authors such as Milton, Tolstoy and Dickens.

The Canadian government, in the form of the Royal Canadian Mounted Police, was responsible for investigations in the north, although these were generally for administrative and political purposes. On the other side of the Arctic, the Soviet Union declared the islands to the north closed to other countries, while conducting a long-term series of hydrographic surveys of the Northeast Passage (now called the Northern Sea Route).

There was more of a scramble in the far south. One of the first major political efforts after the Heroic Age was the British, Australian, New Zealand Antarctic Research Expedition of 1929–1931, led by Douglas Mawson. Although sent to solidify claims to regions that eventually became the Australian Antarctic Territory, Mawson also engaged in extensive scientific research.

While not an official government expedition, the British Graham Land Expedition (1934–1937) was perhaps the most cost-effective effort in the history of Antarctica. Led by Australian John Rymill, it brought back rich results in biology, geology and meteorology, and also established beyond question that the Antarctic Peninsula was a peninsula and not an archipelago.

BELOW The US Coast Guard icebreaker *Northwind* clears the way for USS *Mount Olympus* during the US Navy Antarctic Developments Project (1946–1947), the fourth Antarctic expedition under the command of Byrd.

ABOVE An invoice via the *Expéditions polaires françaises* for the supply of parts for the motor vehicles used on the Norwegian-British-Swedish Antarctic Expedition (1949–1952).

ABOVE Paul-Émile Victor, whose efforts to promote French research in the polar regions resulted in the founding of *Expéditions polaires françaises* in 1947.

be one of mankind's greatest successes in international co-operation, holding all territorial claims in abeyance and setting aside the Antarctic for peaceful, scientific purposes.

In the Arctic, politics and science followed a different course, as the areas north of the countries surrounding the Arctic Basin have long been considered to be under the political – and therefore scientific – control of those countries. In the past few years, this concept has been drawn into question, as a number of countries are increasingly intent on grabbing the natural resources found in or under the Arctic Ocean.

In the years surrounding the Second World War, there were a number of expeditions with varying emphasis between politics and science. These included the German "Schwabenland" expedition (1938–1939), Byrd's US Antarctic Service Expedition (1939–1941), a series of post-war British expeditions collectively known as "Operation Tabarin" – eventually subsumed by the Falkland Islands Dependencies Survey (FIDS) – and Paul-Émile Victor's initiation of the *Expéditions polaires françaises*. One of the most important expeditions focusing on science was the Norwegian-British-Swedish Antarctic Expedition (1949–1952) under John Giaever, which carried out meteorological, geophysical, glaciological and seismic programmes in Dronning Maud Land.

The greatest scientific effort, however, was the International Geophysical Year (IGY) (1957–1958), a world-wide co-operative research programme for which a dozen nations established 55 Antarctic scientific stations studying a broad range of geophysical subjects.

The success of the IGY was not only relevant in scientific circles, however. Diplomats saw that the scientists had worked in harmony regardless of governmental relations. Thereafter, Brian Roberts of the United Kingdom guided the effort that led to the negotiation of the Antarctic Treaty, which entered into force in 1961. For almost half a century, the Treaty and its subsequent and supporting agreements known collectively as the Antarctic Treaty System have proven to

One of the last great polar journeys was made in 1968–1969, when Wally Herbert and three companions made the first surface crossing of the Arctic Ocean via the North Pole, reaching 90° north about a year after Ralph Plaisted, using snowmobiles, had made the first confirmed surface attainment of the Pole.

In recent decades, the Arctic and Antarctic have seen a growing influx of adventurous men and women wanting to prove themselves against nature. Amongst the most accomplished of these are Sir Ranulph Fiennes, Peter Hillary, Will Steger, Jean-Louis Etienne, Ann Bancroft, Borge Ousland and Erling Kagge. Although there is now little true exploration to conduct on the surface of the Earth, it is apparent that the polar regions will remain a focus for politics, science and adventure.

TOP RIGHT Meeting at the South Pole. From left: Hillary, Fuchs and Admiral George Dufek, supervisor of US Antarctic programmes.

ABOVE Members of *Expéditions polaires françaises* depart from Rouen in April 1950, heading towards Greenland. The ship was the Norwegian cargo carrier *Hillvaag*.

BELOW Victor Boyarsky uses his tent to help him figure the team's progress during the International Trans-Antarctica Expedition of 1989–1990.

THE FIRST CROSSING OF ANTARCTICA

Four decades after Ernest Shackleton failed to cross the Antarctic continent, the Commonwealth Trans-Antarctic Expedition finally did so. Conceived and led by Vivian Fuchs, the crossing began in November 1957 after extensive aerial reconnaissance and field preparation. From Shackleton Base on the Filchner Ice Shelf, Fuchs continued through extremely hazardous conditions, reaching the South Pole on 19 January 1958. There he was met by Sir Edmund Hillary, who, after establishing depots outward from Ross Island, had continued to the Pole. Fuchs finally reached Scott Base on 2 March 1958, having covered 3,473 kilometres (2,158 miles) in 99 days.

THE INTERNATIONAL TRANS-ANTARCTICA EXPEDITION

In 1986, Will Steger of the United States and Jean-Louis Etienne of France met while making their way to the North Pole. It did not take long for them to establish an audacious plan: crossing the Antarctic continent by its longest axis. In 1989–1990, the scheme came to fruition as the International Trans-Antarctica Expedition. Steger, Etienne, Geoff Somers of the United Kingdom, Victor Boyarsky of the Soviet Union, Keizo Funatsu of Japan and Qin Dahe of China spent 222 days skiing and sledging (with 36 dogs). They covered 5,586 kilometres (3,471 miles), the longest journey ever in Antarctica.

Translation

Page 49 The Amundsen Diary

This is a fairly literal translation. Words in brackets are for clarification, (?) denotes missing / illegible word, ? denotes uncertainty about a word.

74

..... proper tent site. Achieved an excellent height at mid day which gave 88°30' south latitude or 1' off ? (dead) reckoning (or: 1' to chart later ?). It is fine. An excellent azimuth this afternoon.

Monday 11 December.
Nice weather again. Slight SSEasterly breeze and minus 28°. Partly absolutely clear. Partly some passing haze/mist. A large magnificent ring around the sun. The meridian height gave one minute less than the reckoning. Made up for it by (travelling?) 17 instead of 16 km. Now positioned at 88°56' southern latitude. Conditions (on the ground) and terrain (still) the same. We can certainly notice that it is much harder to work up here at altitude (in the high terrain). We get short of breath by even just (?) – yes, but I suppose they will manage. We look forward to at some point (in time) getting back down to normal altitude.

Tuesday 12 December
Magnificent weather, almost calm and partly clear. Approx. minus 25°. The same good/nice terrain and conditions (on the ground). On the hypsometer it looks like ? we do downwards – but fairly slight. Possibly it is just the weather conditions have this effect. The midday observations – which were made during the most favourable weather conditions – calm, clear

75

sharp sun and equally (sharp, clear) horizon – gave 89°6'. Observations and reckoning are thus again totally in agreement. We have made our usual 17 km and are now positioned on 89°15' – 3 days march from our goal.

Wednesday 13 December
Our best/nicest day up here. It's been calm most of the day – with strong/burning sunshine. Conditions (on the ground) and terrain have been the same. The crust on the snow is fortunately so hard that it is only the slightest amount the sleds and (?) sink in. The hypsometer is still showing decrease even though quite slight so one may assume that we have not already got the highest plateau (always?) but also falling down towards the other side. We have made 15 km today and are now positioned at 89°30'30" according to the midday observation. The observation and the reckoning are in brilliant agreement every day. We can only rely on our only sextant – the Fram sextant - the other one was unfortunately subjected to an impact and has proved not to be reliable. H(elmer) H(ansen), W(isting) and I are now using the Fram sextant. The same fine weather all morning.

Thursday 14 December
It got cloudy after we had got the altitude and arrived with snow showers from SE. The meridian altitude as well as the mercury glass gave 89°37'. The reckoning gave 89°38.5'. This is of course excellent. Minus 23° all day.

76

Afterwards we have driven 8 km and are now positioned 15 km from the pole.

Friday 15 December (actually the 14th)
In the end we reached our goal and planted our flag on the geographic South Pole – King Håkon VII's plateau. Thank God. It was 3 o'clock in the afternoon when this took place. The weather was the best (?) when we started off this morning, but at 10 o'clock (?) it started getting cloudy and hid the sun. A fresh breeze from SE direction. The ground conditions have in part been good, and in part poor. The plateau – the King Håkon VII plateau – has had the same view – fairly flat and without what one may call (?). The sun appeared again in the afternoon and we shall go out to get the midnight altitude. Obviously we are not positioned on exactly 90°, but we must, after all our excellent observations and reckonings, be pretty close. We arrived here with 3 sledges and 17 dogs. H(elmer) H(ansen) slaughtered one just after we arrived. "Helge" had (?). Tomorrow we will go out in 3 directions to encircle the polar area. We have consumed our festive meal. A small piece of seal meat for each one. We will leave from here the day after tomorrow with 2 sledges.

77

The third one will be left here. We are similarly leaving a small 3 man tent (Rønne ?) with the Norwegian flag and a pennant with "Fram" written on it.

Saturday 16 December
A very (?) day. We went out at 12 midnight to take/get an altitude. This we got. The calculations gave us approx. 89°56'. That didn't look too bad. At 2.30 in the morning Bjaaland, Wisting and Hassel went out skiing in order to encircle the Pole. Bjaaland continued (along) our original course NE / N (compass) while Hassel went out in the NW / W and Wisting in SE / E (compass). They were supposed to go a distance of approx. 10 km. Each man carried with him a pole (banner) with a flag on. Our pole had a small bag tied to it, containing an account of where our "Polheim" (polar home) was situated. It was (?) weather. Calm / somewhat misty. H(elmer) H(ansen) and I stayed behind to do observations. Our intention was to do observations every 6th hour. At 6 o'clock in the morning we did the first observation after having taken the midnight altitude. To my surprise this showed a lower altitude than the assumed altitude at midnight. We had clearly during our march from 88° S latitude got quite some distance away from our meridian. The fact of the matter was that we (?)

78

because ? at this latitude it was difficult to ascertain the azimuth. So now it was important to find out which meridian we were on. We then started hourly observations with glass and mercury horizon. Fortunately, the weather remained good and allowed us to get the most excellent observations. Glass and mercury (?) agreed with each other. Because between 5 and 5.30 in the afternoon ? ? Framheim – we found the meridian. This showed us that we had arrived close to the 123rd meridian longitude 0 latitude – not that strange at these latitudes, when one could not determine one's direction. The meridian observations gave us 89°54'38" S latitude and the position of the Pole after compass bearing/direction finding in NW ¼ W. This was an invaluable day and well worth the bother. We ? now with fairly great exactness to find the point of the pole. At 10 in the morning the 3 (men) returned having carried out the task well. The weather has been brilliant all day. The two sledges we are continuing with are now standing ready and if the third ? distance ? no. 2 will be left behind.

79

Helgi (?) sons(s)? was (were?) killed yesterday and in greediness devoured by his comrades. We are departing tomorrow towards the pole point 5.5 km from here. We now have food for 18 days for us humans. The lead dogs for 10 (days). I think things should go well down to our depot at 88°25' and from there to the depot under "The Devil's Glacier". It is quite interesting to see the sun wandering around the sky at more or less the same height day and night. I would think we are the first (people) to see this strange sight.

Sunday 17th December
Again a strange day. We broke camp this morning and started out for the pole point. Bj(aaland)s sled was left behind, his 6 dogs divided between H(elmer) H(ansen) and W(isting) and he looks (?) what transpires. I myself followed to see how he managed to keep the direction. The weather was still ?? A slight breeze from W / 23°. Absolutely clear and a burning sun high in the sky. It was a pleasure to see B(jaaland) keep the direction. He went ahead as if he had marked out a line he could follow. At 11 o'clock in the morning we had covered the 5.5 km, made a halt and put up our tents.

Further Reading

Books

Ayres, Philip. 1999. *Mawson: a Life*. Melbourne: Miegunyah Press, Melbourne University Press.

Baughman, T H 1999. *Pilgrims on the Ice: Robert Falcon Scott's First Antarctic Expedition*. Lincoln, NE, and London: University of Nebraska Press.

Bomann-Larsen, Tor. 2006. *Roald Amundsen*. Sroud, UK: Sutton Publishing.

Bryce, Robert M. 1997. *Cook & Peary: the Polar Controversy, Resolved*. Mechanicsburg, PA: Stackpole Books.

Cyriax, Richard J. 1939. *Sir John Franklin's Last Arctic Expedition*. London: Methuen & Co.

Filchner, Wilhelm. 1994. *To the Sixth Continent: the Second German South Polar Expedition*. Translated by William Barr. Bluntisham: Bluntisham Books, Banham: The Erskine Press.

Fleming, Fergus. 1998. *Barrow's Boys*. London: Granta Books.

Fogg, G E 1992. *A History of Antarctic Science*. Cambridge: Cambridge University Press.

Gurney, Alan. 2000. *The Race to the White Continent*. New York and London: W W Norton.

Guttridge, Leonard F. 1986. *Icebound: the Jeannette Expedition's Quest for the North Pole*. Annapolis, MD: Naval Institute Press.

Hayes, J. Gordon. 1928. *Antarctica: a Treatise on the Southern Continent*. London: The Richards Press.

Holland, Clive. 1994. *Arctic Exploration and Development c 500 BC to 1915: an Encyclopedia*. New York and London: Garland Publishing.

Howgego, Raymond John. 2004. *Encyclopedia of Exploration 1800 to 1850*. Sydney: Hordern House.

Howgego, Raymond John. 2006. *Encyclopedia of Exploration 1850 to 1940: the Oceans, Islands and Polar Regions*. Sydney: Hordern House.

Huntford, Roland. 1985. *Shackleton*. London: Hodder and Stoughton.

Huntford, Roland. 1997. *Nansen: the Explorer as Hero*. London: Duckworth.

Keay, John (editor). 1991. *The Royal Geographical Society History of World Exploration*. London: Hamlyn.

Levere, Trevor. 1993. *Science and the Canadian Arctic: a Century of Exploration, 1818–1918*. Cambridge: Cambridge University Press.

Liljequist, Gosta H. 1993. *High Latitudes: a History of Swedish Polar Travels*. Stockholm: Swedish Polar Research Secretariat.

Mawson, Douglas. 1915. *The Home of the Blizzard*. 2 vols. London: Heinemann.

Nansen, Fridtjof. 1897. *Farthest North*. 2 vols. Westminster: Archibald Constable and Company.

Payer, Julius. 1876. *New Lands Within the Arctic Circle*. 2 vols. London: Macmillan and Co.

Philbrick, Nathaniel. 2003. *Sea of Glory: the Epic South Seas Expedition 1838–42*. London: Harper Collins.

Quartermain, L.B. 1967. *South to the Pole*. London: Oxford University Press.

Riffenburgh, Beau. 2004. *Nimrod*. London and New York: Bloomsbury.

Riffenburgh, Beau (editor). 2006. *Encyclopedia of the Antarctic*. 2 vols. New York: Routledge.

Ross, M J 1994. *Polar Pioneers: John Ross and James Clark Ross*. Montreal and Kingston: McGill-Queen's University Press.

Savours, Ann. 1999. *The Search for the North West Passage*. London: Chatham Publishing.

Scott, Robert Falcon. 1913. *Scott's Last Expedition*. 2 vols. London: Smith, Elder & Company.

Wallace, Hugh. 1980. *The Navy, the Company, and Richard King*. Montreal and Kingston: McGill-Queen's University Press.

Periodical

Polar Record. Published by the Scott Polar Research Institute and Cambridge University Press since 1931.

Websites

Royal Geographical Society
www.rgs.org

Scott Polar Research Institute
www.spri.cam.ac.uk

Acknowledgements

The publishers would like to thank the following people for their valuable assistance in the preparation of this book:

British Library: Peter Robinson
The Ohio State University Archive: Laura Kissel
Institut océanographique, Fondation Albert Ier de Monaco: Muriel Gout
The National Library of Norway: Karen Arup Seip
Mawson Centre: Mark Pharaoh
The Royal Geographical Society: Alasdair Macleod, Jamie Owen, Sarah Strong

INDEX

Figures in *italics* denote illustrations

AAE *see* Australasian Antarctic Expedition
Abbott, George *47*
Adams, Jameson 42, *42, 43*
Adare, Cape 28, 29, 46, 47
aerial exploration 24, 25, 56–57
aerostat 24 *see also* Balloons
airship 56
Albanov, Valerian 51
Aldrich, Pelham 12
Alert, HMS 12, *12*
American Expeditions 6, 16–17, 18–19, 31, 44–45, 56–57
Amundsen, Roald 26, *26*, 40–41, *40, 41*, 46, 47, 48–49, *49*, 56–57, *56*
Amundsen, Roald, diary extract 48–49
Andersson, Johan Gunnar 34, 35
Andrée, Salomon August 23, 24–25
Antactic Explorations 6–7, 26–29, 32–39, 42–43, 46–59
Antarctic 34–35, *34*
Antarctic Continent
 crossing attempts 50–51, 54–55
 first sighting 6
 first on 29
Antarctic Peninsula 29, 58
Antarctic Sound 34
Antarctic Treaty System 59
Archer, Colin 22
Arctic Explorations 2–25, 30–31, 40–41, 44–45, 56–59
Arctowski Henryk 26, *26*
Argentina 34–35, 37
Armitage, Albert 33, *33*
Armytage, Bertram 42
Atkinson, Edward *47*
Aurora 52, 54, 55

Australasian Antarctic Expedition (AAE) 52–53
Australian Antarctic Territory 58
Austrian Expeditions 10–11
Austro-Hungarian Exploring Expedition 10

Back, George 6
Bage, Robert 52
Balaena 37
balloons 23, 24–25, 32, 36, 50
Barrill, Ed 45
Barrow, John 6
Bartlett, Bob 44
Bay of Whales 48
Bay, Edvard 30
Beardmore Glacier 42, 47, 54
Beardmore, William 42
Beaumont, Lewis 12
Beechey, Frederick 6
Belgian Expeditions 26–27
Belgica 26–27, *26*, 40
Bennett Jr, James Gordon 16, 17, *17*
Bernacchi, Louis 28, *28*
Bessels, Emile 8, *8*
Bjaaland, Olav 49, *49*
Bjornen 50
Blisset, Arthur *33*
Booth Island 38
Borchgrevink, Carsten 28, 29
Bowers, Henry ("Birdie") 47, *47*
Boyarsky, Victor 59, *59*
Brainard, David 18, 19
British Arctic Expedition 12–13
British Expeditions 6, 12–13, 28–29
British Graham Land Expedition 58
British National Arctic Expedition 29

British, Australian, New Zealand Antarctic Research Expedition 58
Browne, Belmore 45
Browning, Frank *47*
Bruce, William Spiers 27, 37, 54
Brusilov, Georgi 51
Buddington, Sidney 8
Bull, Henrik 27, *27*
Byrd, Richard E 56–57, *56, 57*, 59

Cagni, Umberto 31
Camp Ridley 28, *28*, 29
Campbell, Victor 46, *47*
Canadian Arctic 6, 9
Cape Adare 28, 29, 46, 47
Cape Denison 52
Cape Evans 46, 47
Cape Royds 42, 43
Cape Sabine 18, 30
Challenger, HMS Expedition 12, *13*
Charcot, Jean-Baptiste 26, 34, 38–39, *39*
Cherry-Gerrard, Apsley 47, *47*
Chester, H C 8
Christansen, Fred 18
Coates, James 37
Coats, Andrew 37
Colbeck, William 29
Collins, Jerome 16, 17
Commonwealth Trans-Antarctic Expedition 59
Conger, Fort 18
Conway, Martin 25, *25*
Cook, Frederick A 44–45
Cook, Robert A 26
Crean, Tom 55
Cross, William 19

Dahe, Qin 59
Dallmann, Eduard 39, *39*
Danskøya 23, 24
Dash Patrol 49, *49*
David, T W Edgeworth 42, *42*, 43, *43*, 49

Davis, John 29
Davis, John King 52, 53, *53*, 55
De Gerlache, Adrien 26, *26*
De Long, George Washington 16, *16*, 17
Denison, Cape 52
Denman Glacier 53
Deutschland 50–51
di Savoia, Luigi Amedeo (Duke of Abruzzi) 31, *31*
Dickason, Harry *47*
Dickson, Oscar 14, 15, 24
dirigible 25
Discovery 28, 32, *32*
Discovery, HMS 12
Dobrowolski, Antoni 26
dogs 21, 22, 28, 30–31, 32, 40–41, 43, 46, 48–49, 50–51, 52, *54*, 59
Drygalski, Erich von 36, *36*, 53
Dufferin, Lord 7
Duke of Abruzzi *see* di Savoia, Luigi Amedeo
Dumont d'Urville, Jules-Sébastian-César 6
Duse, Samuel 34

Eastern Party 46
Eckholm, Nils 24
Eielson, Ben 57
Elephant Island 54, 55, *55*
Elison, Joseph 18, 19
Ellesmere Island 18, 30, 44
Ellsworth, Lincoln 56–7, *56*
Endurance 54, *54*
Erebus, Mount 42
Etienne, Jean-Loius 59
Evans, Cape 46, 47
Evans, Edgar 33, 47, *47*
Evans, Teddy 47
Expéditions polaires françaises 59, *59*

Falkland Islands Dependencies Survey (FIDS) 59
Ferrar Glacier 33
Fiala, Anthony 41

Fiala-Ziegler Expedition 41
FIDS *see* Falkland Islands Dependencies Survey
Fielden, H W: extract from diary 12–13
Filchner Ice Shelf 50, 59
Filchner, Wilhelm 50–51, *51*, 54
flying boats 56
Fort Conger 18
Frænkel, Knut 24
Fram 22–23, *22*, 30–31, *30, 48*, 48–49
Fram Expeditions
 First 22–23
 Second 30–31
 Third 48–49
Fram, plan of 22–23
Framheim *48*, 49
Français 38, *38*
Franklin Expedition 6 17, 40
Franklin, John 6, 7, 40
Franz Josef Land 10, 11, 22, 37, 51
fraudulent claims 44–45, 56
French Expeditions 6, 38–39
Fuchs, Vivian 59, *59*
Funatsu, Keizo 59

Gamél, Augustin 21
Garwood, E J 25
Gauss 36, *36*
Gaussberg 36, 53
German Expeditions 36–37, 50–51, 58
German South Polar Expedition 36–37
Gjøa 40–41, *40*
Gjøahavn 40
Graham Land 38
Great Ice Barrier 6, 29, 32, 42, 46, 48, 54
 see also Ross Ice Shelf
Greely, Adolphus W 18, *18, 19*, 19, *19*, 30
Greenland 8–9, 18–19, 20–21, 31, 40
Gregory, J W 25
Grunden, Toralf 34
Guilder, William Henry 17, *17*

Gyldén, Olof 35

Hall Island 10
Hall, Charles Francis 8, 9, *8, 9*
Hamilton-Temple-Blackwood, Frederick *see* Dufferin, Lord
Hansen, Godfred 40
Hanson, Nikolai 28
Hanssen, Helmer 40, *41*, 48
Harmsworth, Alfred 23
Hassel, Sverre 49, *49*
Hearne, Samuel 6
Heiberg, Axel 30
Henry, Charles 19
Henson, Matthew 44, *44*
Herald *see* New York Herald
Herbert, Wally 59
Heroic Age, the 8, 56
 prelude to 6–7
Hillary, Edmund 59, *59*
Hollick-Kenyon, Herbert 57
Hooper, William H extract from journal of 6–7
Hope Bay 34–35
Hudson's Bay Company 6
Hurley, Frank 52, *54*
Hut Point 32, 33, 43

Imperial Trans-Antarctic Expedition (ITEA) 53, 54–55
International Geographical Congress, Sixth 26, 27
International Geophysical Year 59
International Polar Year 19
International Trans-Antarctica Expedition 59
Inuit, Netsilik 40
Investigator 6, 7
Irizar, Julián 35
Italia 57
Italian Expeditions 31

Jackson, Frederick 23, *23*, 43
Jackson, John 17

Jackson-Harmsworth Expedition 23, 37
James Caird 55, *55*
Japanese Expedition 49
Jason 21
Jeannette, USS 16, *16*, 22
Jeffryes, Sydney 53
Jensen, Berhard 29
Johansen, Hjalmar 22, *22*, 49, *49*
Jones, Sydney 53

Kaiser Franz Josefs Land *see* Franz Josef Land
kayaks 22, 40
King Edward VII Land 49
Kling, Alfred *50*, 51
Koldewey, Karl 11, *11*
Kondrat, Aleksandr 51
König, Felix 51
Kuznetsov, Aleksandr 57

Lamont, James 7, *7*
Larsen, Carl 27, 34–35
Larsen Ice Shelf 34
Lashly, Bill 33
Lena 14, 15
Little America Times 56–57
Lockwood, James 18, 19
Lorenz, Wilhelm 51

Mackay, Alistair 42, *42*
Mackenzie, Alexander 6
Macquarie Island 52
Madigan, Cecil 52
Main Base 52
maps
 Antactic 33, 35, 38, 51, 53, 57
 Arctic 7, 9, 11, 15, 17, 23, 42
Markham, Albert Hastings 12, 13, 18
Markham, Clements 26, 28, 29, 32, 33, *33*
Marshall, Eric 42, *43*
Maud 41
Mawson, Douglas 42, *42, 43*, 52–53, 58
McClintock, Francis Leopold 7